About the Author

Tomm Stanley's life experience has taken him from his birth town of Milwaukee, Wisconsin, through to relocation by his family to Orlando, Florida, overseas work in Antarctica, New Zealand, Australia and South East Asia - finally to his own relocation to live in New Zealand.

Tomm's early work experience in inventory management provided him with the skills that unlocked a door into Antarctica, an experience he remembers fondly and a place that he claims changed his life. Working within the area of Logistics and Materials Supply for the US Antarctic Program's operations support contractor, between 1991 and 1994 he spent 28 months in the Antarctic. During this period he learned of the basic concepts of solar design and produced several experimental devices to satisfy his curiosity.

In 1994, on the way back to America from his second trip to the Antarctic he had a chance meeting at a skydiving drop zone (Tomm was involved in the sport for many years) that led to an offer to work in New Zealand. Two years after arriving in the country, with a grant of permanent residence by the New Zealand Immigration Service, Tomm settled in a remote valley on Banks Peninsula, not far from Christchurch, the largest city on New Zealand's South Island. On this 150 acre property, Tomm began an initiative to re-forest most of the property's steep hillsides back to indigenous forest species. He set up a native tree nursery and a peripheral web-based business, www.heartbeatnursery.co.nz to support the planting project's reforestation activities.

In 2000, he began a new career path as a public speaker, corporate trainer and group facilitator. Initially working within the motor trades allowed him to establish himself quickly as an expert due to his experience within that industry. He has also facilitated meetings with local community groups and committees preparing to make submissions to the International Antarctic Treaty.

Tomm has sought educational experiences in all places; universities, competency based training, recognition of prior learning, libraries and wherever reading material offers a new piece of knowledge. With personal interests ranging from music and arts to physical sciences and ecology, he has put together a very special combination of skills, knowledge and hands-on experience.

Tomm Stanley is available for public appearances and training workshops covering a wide range of ecology, management and service related topics worldwide. He may be contacted via e-mail at the website www.heartbeatnursery.co.nz.

Planting for the Future

Going Solar

I would like to dedicate this book
to one of my longest standing friendships
and the individual that, more than anyone else,
opened my mind to the wonders of
the physical sciences, Brian Mills.
Believe it now, I really was listening.

Other Works by the Author

Stone House – A Guide to Self-Building with Slipforms

ISBN 047309970-5

Going Solar

Understanding and Using the Warmth in Sunlight

Written and Illustrated by
Tomm Stanley

Photography by Tomm Stanley
unless otherwise indicated*

*See historic image statement on next page

Published by
Stonefield Publishing

First Published 2004

Cover design and Title Concept: Matt Searles of Red Shed Design.

ISBN 0-476-01082-9

Acknowledgements

I would like to extend a grateful acknowledgement to the following individuals for their assistance in checking facts, suggesting ideas for improvement and other elements that supported the creation of this book:

Michelle Finnemore, Dr. Michael Finnemore, Matt Searles, Mark Walker, Sabrina

Historic Images

Printed for Stonefield Publishing, PO. Box 55, Tai Tapu, Christchurch 8150, New Zealand

Pope Print Christchurch & Timaru.

Table Of Contents

Foreword   1

Introduction   9

Section One – History and the Basic Principles of Solar Design

Chapter One – The Basic Principles of Solar Thermal Design   13

Chapter Two – A History of Solar Powered Invention   41

Section Two – The Science Behind the Sunlight

Chapter Three – It All Starts at the Sun   67

Chapter Four – Nuclear Science Basics   79

Chapter Five – Principles of Heat in the Universe   85

Section Three – The Nature of Materials

Chapter Six – Materials   99

    Facts About: Thermal Storage   101

    Facts About: Insulators   107

    Facts About: The Role of Color   117

    Facts About: Heat Absorption and Emission   121

    Facts About: Selective Surfaces   123

    Facts About: Glass   127

Section Four – Practical Applications

Chapter Seven – Application of the Knowledge   131

    Study One: Sun Tea   133

    Study Two: Parabolic Cooker Experiments   135

    Study Three: Solar Thermal Electricity   140

    Study Four: Solar Ovens   141

    Study Five: The Solar Toilet   143

    Study Six: Solar Tracking   147

    Study Seven: Passive Solar Construction   151

Finale

    169

Appendices

Appendix One – A Chronology of Solar Design Achievement   170

Appendix Two – IAU Nomenclature   181

Appendix Three – Resources   182

Glossary

    183

Index

    195

The Beginnings of My Interest

My interest in alternative energy and self-sufficient lifestyles started in 1993 while working the second of two contracts with the U.S. Antarctic Program. I was supervisor for the materials operations that supplied the McMurdo and South Pole Station construction crews. The construction crews were a divided group at that time. There was the General Construction crew, which handled the typical maintenance and worked on new buildings in and around McMurdo Station. There was the South Pole Construction crew, which was the same as general construction at McMurdo, only they worked down at the Geographic South Pole building a new station that was partly replacing and partly adding-to the station that had been in operation there since the early seventies. Finally there was the Science Carpenters who worked out at the remote research sites. It was this latter group that had the strongest influence over my growing interest in alternative energy and building techniques.

A composite panorama of McMurdo Station, Antarctica. The presence of the U.S. Coastguard Cutter at the ice pier (left centre) would place this series in mid-Austral Summer, perhaps December or early January.

Quite a few of the tradesmen in the Science Carpenter's crew were what could be called 'greenies'. In their workshop office was a catalogue of alternative energy supplies that was referenced for the different remote site applications that they dealt with. The catalogue had all kinds of things that were then new and fascinating to me - information on photovoltaics, wind turbines, hydroelectric units, etc. It was after a few good reads through this catalogue that I became seriously interested in alternative energy and wanted to learn more.

My casual curiosity turned into more of a research project and I began scouring through the piles of Popular Science, Popular Mechanics, Discover and other magazines, that were in all the lounge areas around the station. McMurdo also has a pretty good library that revealed several sources on the subjects I was chasing after.

I spent the next several months looking into any publication that might have held the promise of containing any information on alternative energy topics. It wasn't very long before I began to build the "Energy File", a compiled record of every article I came upon in regard to alternative energy and self-sufficient homes (I still have the Energy File and am always adding bits to it as I find them). All this research got my brain turning and it wasn't long until I decided to have a play with some of these ideas. After all, through working with the carpenters I had two complete workshops at my disposal and a nearly unlimited supply of construction materials. All I needed to do was spend a bit of time designing something and then build it.

Photo credit: Anonymous

My "Hero Shot" as we called photos at significant landmarks; this one at the ceremonial pole, South Pole Station, Antarctica.

The station's central dome is in the background.

Designing wasn't so difficult but building was another story. There wasn't a lot of time for personal projects in the Antarctic as we worked a 54-hour week over 6 days. It may seem long to the average 'real world' work ethic-tuned mind but there isn't a lot of opportunity for outside activities down there anyway. Work keeps you busy and that keeps your mind healthy. The summer season ended too soon and before I had anything related to alternative energy completed to a testing stage, we were facing the winter season and 24-hour a day darkness. It was during this time that I met and became friends with the station's machinist, Marc.

Marc also had a bent for alternative dwellings and had a series of books called *Earthship* that had been sent to him. This stuff really impressed me for much of what was in the Energy File was represented here in these homes. For those that have not encountered Earthships, their exterior and loadbearing interior walls are built of used automobile tires that are rammed full of earth and stacked like bricks. Aluminium cans set in a concrete mortar are used for non-structural interior walls and toward the end of the construction, both the tire and the can walls are covered in a concrete plaster so as to create a very pleasant, earthy-looking structure.

The rammed earth tires that made the exterior walls of the Earthships just seemed too far out to me. In morning and afternoon break times with the carpenters I brought up Earthships and how they were constructed as a topic of conversation. I actually got the attention of a few of them, not really in a positive way for the most part, and so I took a bit of ribbing along the way from the 'doubters' in regard to my interests. I actually rang one of the doubting guys that was also from Florida about a year later when we were back home. I had *him* on about it, giving him a call and saying that I was ready to start building and wondered if he was interested in coming over to Orlando to give me a hand ramming earth into tires.

Silence on the other end of the phone…

Until I said "Nah, I'm just kidding!"

It was good for a laugh but he wasn't interested.

Foreword

Getting back to Antarctica, with the Earthship concept fuelling my interest even further, I designed and built a small solar oven and began the wait for the Sun to show itself again. The oven had about 1 cubic foot capacity with a hinged top for easy access. It was made of 5/8-inch (15mm) plywood with drywall for an inner lining. The idea here was that the drywall would act as thermal mass to help keep temperatures up. This was further lined with sheet metal that was painted black. On the outside was a layer of 1-inch extruded polystyrene insulation (blueboard) and 1/4-inch (6mm) Lexan for glazing. I had a probe-type thermometer, that was designed to work in a boiler, screwed into the side of the oven to monitor the inside temperature.

Summer season finally did arrive and before long the little oven was out on the back deck of the carpentry shop, getting the first doses of sunlight that any of us had seen in a long time. I checked the temperature on it every couple of hours hoping that the thing would work. Hoping, because I wasn't the only one watching it and I really wanted to say "I told ya' so" to some of the doubters. The temperature in the oven came up but not much. The doubters were winning but it was still early days in the testing program. I decided to add a bit more insulation, giving it two inches of blue board and also added a second layer of glazing. After the upgrades it was back to the deck and I was praying for better results this time.

Falling off the bottom of the Earth!

Having some fun at the Geographic South Pole marker .

Photo credit: Anonymous

Results of solar testing in the Antarctic can be rather slow going as the good weather will pack up and leave for extended periods of time. This gives the doubters plenty of opportunity to come up with new digs, jokes and comments. However, it also gives the supporters time for more support and suggested modifications. For people that may have an experimenting habit, and are brave enough to share it with their friends, they may know that there's nothing quite like suggestions from well meaning friends and acquaintances. I have always believed, and the situation with the solar oven reinforced the beliefs, in the mechanic's theory of labor charges:

- Labor costs $20.00 per hour;

- $30.00 per hour if you want to watch;

- $40.00 per hour if you want to help.

That said, there were some keenly interested people that really wanted to see the little oven work well, and they too were learning about solar power from it. Supporters and doubters alike had fun with it and when you're on the bottom of the world, fun is important.

Eventually the Sun did come back out and the temperature in the little solar oven rose to acceptable levels. It was over 100°F (38°C) inside the oven this time around. That was with an outside temperature of around minus 30°F (minus 35°C), so the little oven was doing some serious heating. The doubters stopped doubting. This was really cool! I brought it back in for one more round of modifications. Four inches of insulation this time, a complete exterior shell of 5/8-inch (15mm) plywood and a third glazing; then it went back out to the deck.

The temperature again improved and the little oven was gaining interest. I had a feeling of success with this project when the crew started watching the thermometer, and my team in Materials Supply began issuing updates as they would come in for breaks or between material delivery runs. The top temperature recorded was 115°F (45°C). I played with the glazing a little more and found little difference between two and three layers but decided three was probably the best bet for the extreme cold we were dealing with. I felt a great sense of achievement and didn't feel the need to do any more fine tuning to extract even better performance, the most important thing to me at this point was that *it worked.*

After this small victory with the little oven I decided to build something that would actually be of use. I decided on a passive solar heating unit that could be framed into the wall of a building. Installation would be just as if framing in a window and it was therefore dubbed the "Solar Window". The Solar Window was basically a miniature Trombe Wall* with a frame around it. Made of 1x6, 1x2, drywall, sheet metal and three layers of Lexan it was rather on the heavy side. The weight was good though as it was the result of the drywall, which was functioning as the unit's thermal mass.

The Solar Window was amazing. I had the attention of the Science Carpenter's Co-Ordinator, Woody, on this one. Again the back deck of the carpenter shop was the test site. Outdoor air temperatures were up to around minus 10°F (minus 23°C) by this time and the weather was improving. This made the testing go somewhat quicker and there were actually no modifications made. The Solar Window cranked out hot air around 110°F (42°C) while standing out in the cold with only a 1-inch (25mm) piece of blue board covering the side that would normally be open into the room. Woody decided to give the unit a go out in the field and arranged to install it in a building at a research site somewhere in the middle of nowhere.

* Trombe walls are discussed in detail later in this book.

Unfortunately for me, my contract with the Antarctic Program was up and I had to leave before the Solar Window was installed. At the time, I was looking forward to returning to the Antarctic the following year and intended to pursue working for the Science Carpenters. Other opportunities arose for me however. A job offer from a company in New Zealand, and my acceptance of the offer, changed my plans for continued involvement with the U.S. Antarctic Program. I contacted Woody some months later when he was back to the real world to find out how well the window had worked in actual field use. Sadly, I was told that it had been removed from its scheduled flight out to the research site and before being rescheduled and finally sent, it was broken in handling at the helicopter pad and thrown away.

Foreword

Introduction

Luckily for us, the Sun shines every single moment of every single day, and every minute of every day there is sunshine somewhere in the world. Lots of sunshine reaches our planet every day. Consider for a moment that sunshine contains useable energy, lots of useable energy. Science tells us that over four zillion kilowatt-hours of energy comes to Earth from the Sun each day. That's enough energy to run entire cities on.

In my opinion and that of many other people looking for a sustainable future, this world needs a source of energy that will last until humans no longer exist and will keep our planet a pleasant place to live in the meantime. It would be great too if people didn't have to pay for the energy that they need to survive on a day-to-day basis; staying warm, cooking food and having a generally happy existence. Fossil fuels, our world's current main source of power, just don't make the grade. From a polluting point of view we know they are a disaster on a global scale. I am not suggesting that people reject the use of petrochemicals full stop; where would we be without oil-based lubricants, plastics and all the other synthetic products produced from them? My objections come when there are reasonable alternatives that meet the needs of people as well as petrochemicals, in a non-polluting fashion. Interestingly, fossil fuels are the remnants of sunlight from days gone past, the remains of forests and animals from the prehistoric era. The reality is, there are alternatives to consuming fossilised heat, to create heat. Enter **solar thermal** design; the practice of designing devices to collect heat from the Sun.

For a bit of background on how this book evolved, my interest in most things includes having a deeper understandings of how they work and why they are the way they are. When exploring solar designing, one of my early questions about sunlight was "yes it's warm, but *why* is it warm?"

And in regard to the way materials warm when contacted by sunlight; again, "Why?" and also "How?"

While most every solar designing book that I have come across tells readers to put a dark surface in front of sunlight to make it warm, the reasons for this phenomenon – the science involved and the physical properties of sunlight and materials – do not seem to be an area very well explored in the world of practical, everyday, hands-on people. I had to spend quite a lot of time searching through physics books to find the answers I was looking for and while I think that science is an awkward topic for a lot of people, I find that the way in which science is written about doesn't usually help that issue.

I consider myself fortunate to have among my friends, scientists and researchers whose interests range over a variety of studies. The science content that I have written about here had to be reviewed (several times) by these individuals before I could consider it finished. Condensing miles of information into a few feet of printed page has presented its challenges and the goal with the science was to make it enjoyable to read and still pass on the relevant information. It seems that while reviews from some hands-on people indicate that portions of the science content contained here is a little 'heavy', the scientists think it's too condensed. I suspect that means I've struck a reasonable balance.

By covering basic solar designing principles, some history in the field of solar thermal devices, getting into the science involved in the solar heating phenomenon and then wrapping up with some practical applications, with this book I am attempting to bridge the gap between scientific theories and practical, everyday, hands-on people. Hopefully it will help people to design devices that can take advantage of solar heat and assist them in truly understanding the "why" and the "how" of what's going on in that warm light and the materials that they are working with.

Solar thermal design's simplest manifestation, **passive solar** design, is hardly an earth shattering revelation. No intervention is required by people or machines, hence the term passive. Stand in the sunlight and you will feel warmer than when you are standing in the shade. Place an object so that the Sun can shine upon it and it will become warm; this is passive solar. The Sun has been used by humans to keep warm for longer than we have kept records and it is speculated that the origins of life on this planet would not have risen from the primordial soup if the Sun was any more that 5% closer or farther away than its current position. I think we should take a cue from that ancient slime pool and make good use of this amazing resource to create something useful for us today, and keep a beautiful planet for those following us tomorrow.

The Basic Principles and History of Solar Design

Chapter One

The Basic Principles of Solar Thermal Design

Solar thermal designers, like designers in all fields of interest, need a basic set of knowledge that can function as both a strong foundation and a source from which to draw inspiration. The best designers in fields such as engineering, architecture, software, chemistry and others generally start with a basic knowledge of the materials and/or systems that they are working with. They will know what has been achieved before and will perhaps be aware of the cutting edge of what is currently being achieved. From this basis and with their own spark of ingenuity, new ideas and systems are often born. It is the rare individual that starts with absolutely no knowledge of an area and breaks new ground.

This highlights one of my own favourite ways to learn, a process known as relational learning. In this manner of self-education, I look into subjects that relate to the main topic of interest that I am studying, a process that often leads my mind to new twists on current information. This learning habit comes through in my writing as well, so as I explore a basic subject, say the Sun, I will touch on other related topics such as the solar system or chemistry or whatever might be applicable. I believe that this method helps to expand my awareness of the world around me and often leads to new learning endeavors when a relational topic becomes a focused interest. Of course this eventually leads to yet more relational topics that spark again newer interests ... on and on it goes; and what fascinating reading all this relational stuff makes as well.

Collection Methods

When considering the different ways that solar heat may be collected, the term **gain** is used in place of the word collection. The three basic types systems used to collect solar heat, and the terms used to describe their respective forms of gain are:

•   Direct gain

•   Indirect gain

•   Isolated gain

Systems may be customized and integrated so that gain-type hybrids are created or particular site conditions may be used to greatest advantage. Most designers will find situations where they can incorporate both direct and indirect gain systems very easily. Isolated gain systems are most often used in special circumstances like solar hot water heaters, where containment is required to be in a particular place, not necessarily convenient to where the collection occurs.

Direct Gain

If the sun is shining directly into a space or onto an object that is to be warmed, then **direct gain** achieves the heating effect. A simple version of direct gain is the unwelcome, blistering hot vinyl seat or steering wheel found in a car that has been parked out in the Sun. The sunlight streams through the window and heats any surface that it happens to fall upon. Direct gain is the simplest form of heat collection that can be used in a solar heated device.

The exposed metal in the steering wheels of some cars is likely to stay hot for an extended period of time, providing us with an example of thermal storage, which is also important for devices collecting solar heat but not so great for drivers of the previously parked car. Thermal storage amounts to nothing more than bodies of dense materials capable of soaking up and holding heat. In solar design, such a material is often referred to as a **mass body**.

Common examples of Direct Gain
solar collection:

Laundry drying on a rack
Sunbathing reptiles
Hot water in a garden hose

Indirect Gain

A device in which a collector is first heated, then transfers the heat to where it will be utilised, is considered to be using an **indirect gain** system. A variety of collectors can be used in an indirect gain system; anything from dense materials like stone or steel to water containment vessels. The collector is placed in a position whereby the Sun strikes one side of the collection unit and the heat is used directly on the other side. A key understanding for indirect gain systems is that the collecting side and the using side are two separate spaces, basically divided by the collector.

Space heating of a room through its exposure to direct gain exterior walls is a typical application of an indirect gain system. A masonry wall may be warmed by exposure to sunlight on the outside of a building with heat transfer to the home's interior occurring by the heated mass radiating warmth to the air and/or other materials within the room.

Indirect gain systems can make use of vents placed in such a position that they take advantage of the natural **thermosiphon** effect that is induced. The term thermosiphon refers to a phenomenon in fluids and gasses within a closed system where they circulate on their own due to the dynamics of heat and cold in a fluid material. Air caught in a cavity between a collecting surface and an exterior panel of glass will become warm. The hot air then rises resulting in a drop in pressure below it, which draws cooler air down to take its place at the lower level within the closed space. If vents are positioned properly the warm air can be directed to literally blow into the adjoining room space, while cold air is subsequently sucked out of the room and into the warming cavity. This arrangement of glass, wall and ducts, shown in the diagram on the next page, is called a Trombe Wall, named for the man credited with its invention, the French scientist, Felix Trombe.

There will be more information on the principles at work in this scenario, as well as the cars steering wheel from the direct gain example, coming up in the next two sections, *The Science Behind the Sunlight* and *The Nature of Materials.*

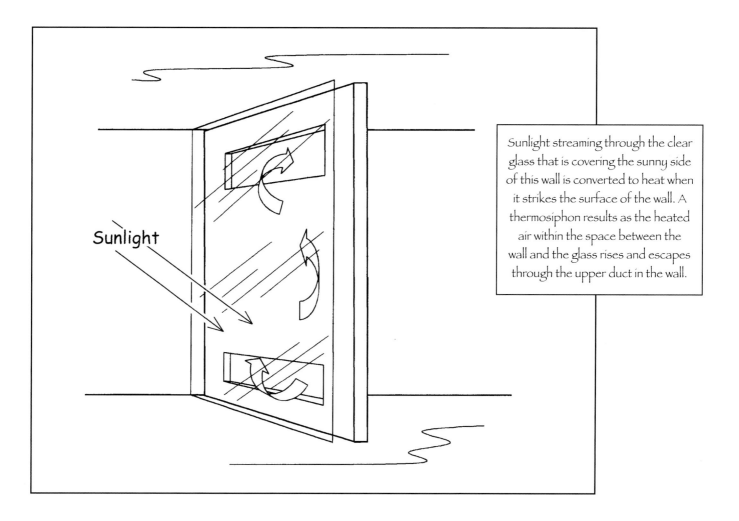

Sunlight

Sunlight streaming through the clear glass that is covering the sunny side of this wall is converted to heat when it strikes the surface of the wall. A thermosiphon results as the heated air within the space between the wall and the glass rises and escapes through the upper duct in the wall.

Isolated Gain

When conditions do not allow the placement of spaces to be heated either directly in the path of sunlight or in a position favorable to the use of an indirect gain system, **isolated gain** systems come to the rescue. These systems gather heat in a collector that is remote from where the heat is being used or stored. Once collected, the warm air or fluid is then transferred via some form of ducting or conduit to the area that requires heating or to the storage material.

Although there are some advantages to using isolated gain devices, notably that systems like indoor solar heating units can be used in areas of a dwelling that have no natural solar exposure, losses of heat during the transfer from the solar collector to the point of use/storage can be observed. Heat losses during transfer can be minimised if **insulation** materials are effectively used along the path that the air or fluid must follow, but by the very nature of materials interacting with their environment, the losses cannot be eliminated completely and efficiency will be lower than with a direct or indirect gain system.

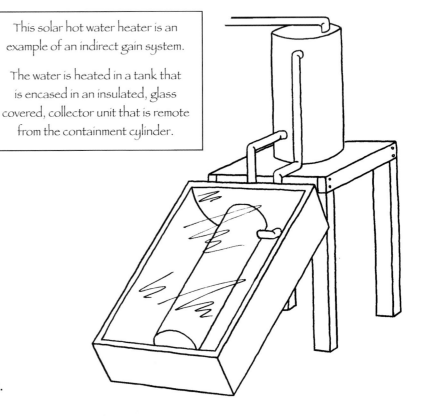

This solar hot water heater is an example of an indirect gain system.

The water is heated in a tank that is encased in an insulated, glass covered, collector unit that is remote from the containment cylinder.

Collector Types

The designer has a wide range of previously explored, field-tested choices to select from when deciding on the type of collector to use in a solar thermal device. As with many things, collectors range from the simple to the complex and the only real constraints when deciding on which option to choose are the actual requirements of the installation and the extent of the construction budget. The following text, and the diagram on the next page, highlights the various options beginning with the simplest and progressing to the more complex.

The simplest collector would probably be nothing more that an exposed collecting body such as a fluid container or mass body that would provide a warming effect in its general vicinity after some amount of solar exposure. We'll cover different material combinations that assist designers in choosing the best containment choices in the *Nature of Materials* section but high tech solutions are not required for an exposed collector to provide a beneficial warming effect.

Flat plate collectors will be found as the next step up from exposed bodies. While still basically a container with a fully exposed surface, flat plate collectors can be designed to warm fluids held within or be ducted into an air-handling system. While typically part of more involved systems that include methods to move the collected heat to its point of use, flat plate collectors may require more thought to their sizing and placement than the simple, exposed body.

The first system that truly requires some form of design work and material fabrication is the plate and tube collector. Consisting of a series of tubes that circulate air, some other gas or a fluid, the adjoining plates function like a heat sink working in reverse. Where heat sinks provide large surface areas to dissipate heat from a central point, the plate and tube collector exposes a large surface area to the sunlight in order to warm the contents of the smaller, central tube.

Significant efficiency gains are obtained by enclosing any of these systems within an insulated container; these systems are called enclosed containment collectors. Via glass or some other transparent surface on one face of the container, this type of collector allows sunlight to strike the collecting surface but through a series of interactions that will be explored in *Section Two - The Science Behind the Sunlight* and *Section Three - The Nature of Materials*, makes it difficult for the captured heat to escape from the container.

Delving into the more advanced systems and those that are less likely to be built by an individual working on their own, evacuated tube collectors take advantage of metal's ability to transfer heat efficiently. By encasing a metal rod in an evacuated (vacuum depleted) glass tube, the rod can obtain high temperatures when exposed to sunlight. One end of the rod will penetrate the glass encasing tube and be integrated into a system that circulates air, gas or fluid to transfer the captured heat to where it will be used or stored.

Some of the most advanced collection systems, and those that generate the highest temperatures, incorporate reflecting surfaces into their design in the way of either a simple reflector that bounces additional sunlight onto a collecting surface or a focusing reflector, which can direct great quantities of sunlight to a relatively small collecting area. In most of these applications, one of the already-mentioned systems will be at the point of reflection or focus to collect the additional sunlight provided by the reflector.

There are a wide range of collector types available to solar thermal designers, from the simple to the more complex.

Reflectors

Reflectors made from a wide range of materials may be used to boost the efficiency of a solar heat collector. Flat, polished steel plates, or better yet stainless steel plates that hold a polish without tarnishing or oxidizing, that are able to bounce additional light onto a collector are likely to be the simplest adaptation of this method. Ranging down the scale of effectiveness, but up the scale for ease of incorporation, aluminum foil is an effective reflector. In the opposite direction, up the effectiveness scale but more difficult to work with, aluminised glass mirrors can be used if the project's budget is adequate to fund such higher-priced options.

Three reflector options shown here are the flat plate, the simple curve and at the lower right, a parabolic trough.

Curving the reflective surface, in order to direct the gathered sunlight to an area smaller in overall surface area than that of the reflector, is a way to intensify the heat found in sunlight The amount of surface area required for a reflector to generate sufficient heat varies with applications. For boiling water in small household quantities, say a gallon (3.8 litres), a reflecting surface with an area of about six square feet (1/2 square metre) focused on a blackened container will do the job in a matter of minutes. Should the reflecting surface have a simple curvature, say a semi-circle, there will be a noticeable difference in the amount of heat gathering on a collector's surface from using a flat panel to simply bounce more light onto the container. If the curved surface incorporates a parabolic shape into its design, that is, a curve capable of focusing the collected light to a precise point, great amounts of heat will occur at that point of focus.

The Sun and Planet Earth

Planning requires information to start with and since solar thermal design is all about working with the Sun, designers working on devices that collect solar heat will need to know about the Sun and especially where it is. Of course the Sun is in the sky but exactly where it is positioned in our sky, at any particular time is the critical consideration.

For those in tune with the seasonal cycles in their area, a person watching the Sun will discover it in a low position in relation to the horizon in the winter, and find it high overhead during the summer. We owe this phenomenon to the fact that our planet orbits the Sun, rotates and is tilted on its rotational axis by 23½ degrees in relation to that orbital path. So remember, the Sun never actually moves, it only appear to move because our planet is orbiting it and rotating. It is both the tilt and orbit of our lovely planet that sets up the conditions for a summer and a winter season.

Planet Earth follows a path that takes it around the Sun once every year. In this orbit, the Earth follows a 'track' that is a rather elongated oval shape called an ellipse; we're held on the track by gravitational attraction between the Earth and the Sun. At some times of the year this track takes our planet significantly farther away from the Sun, the orbital aphelion, than at other times of the year, the orbital perihelion (closest to the Sun).

Johannes Keppler, an astronomer and mathematician born in 1571 discovered this fact and thus explained why the northern winter season is shorter than its summer; the northern hemisphere is tilted away from the Sun (winter) when Earth is closest to the Sun. This just happens to be a position in which the orbit is also fastest because the pull of the Sun's gravity makes the planet race toward it; planet Earth is travelling through **spacetime** at its fastest as it comes around the Sun in the perihelion phase of its orbit. Gravity from the Sun then slows us down as we reach the farthest limit of the elliptical orbit – the aphelion – and the height of the northern summer. Remember, these seasonal considerations are vice-versa for the southern hemisphere.

Spacetime

Spacetime is the fabric of our universe and as inhabitants within this medium we should understand that space and time are one entity.

Space consists of the three dimensions that we are physically aware of (height, width and depth) and possibly others whose existence science has speculated. Time is a measure of intervals between events.

Experimentation has shown that the two exist in an intriguing union – spacetime – rather than as two distinct phenomena.

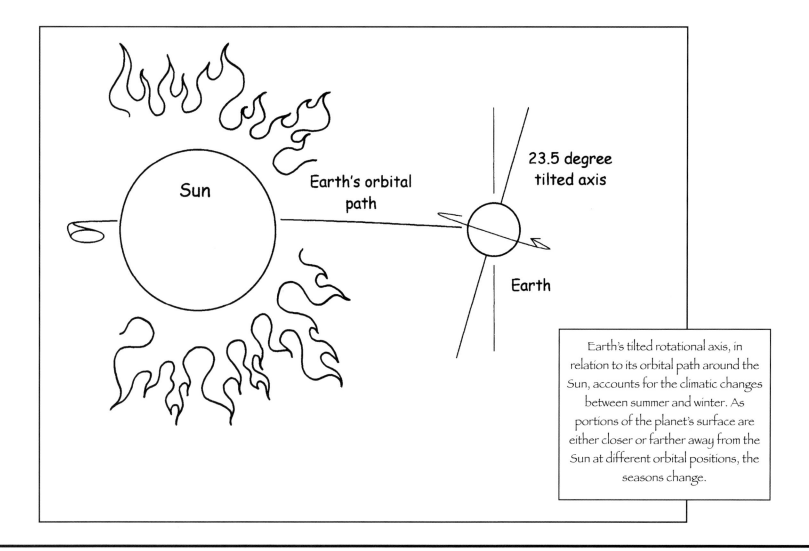

Sun

Earth's orbital
path

23.5 degree
tilted axis

Earth

Earth's tilted rotational axis, in relation to its orbital path around the Sun, accounts for the climatic changes between summer and winter. As portions of the planet's surface are either closer or farther away from the Sun at different orbital positions, the seasons change.

Fixed-Collector Considerations

The seasonal fluctuations are important to consider because many devices will have fixed or inconveniently moved solar collectors. Devices that follow the Sun's path in the sky, an ability known as **solar tracking**, can be rather complex to design and create, so I'll stay focused on fixed collector installations throughout most of the book. However later, in the *Practical Applications* section, I will explore a fairly simple tracking design that most people could duplicate.

In fixed collector situations, designers need to consider what time of the year it is that they want the device to have optimum solar collecting potential and then position the device's collectors to take advantage of the Sun at that point in time. If the device is to be used year round, many designers would aim for maximum collection during the coldest months of winter. This consideration will determine the angle of the collecting surface above zero, or lying flat on the ground; in other words, the devices **vertical orientation**.

On a smaller timeframe, designers must also consider the day-to-day path of the Sun in the sky. We cannot forget that our planet is spinning through space like a top and the very reason that we have a day and night is because as Earth rotates on its axis, our position on its surface comes into view and away from view, of the Sun. From a place low on the eastern horizon early in the morning, somewhere overhead at midday to low on the western horizon just before nightfall, the Sun passes through the sky. Designers need to make a choice; from where they look out and away to the Sun from a chosen site, do they point the solar collecting device to the left, straight ahead or to the right? In other words, what is the device's **horizontal orientation**?

To help with placing solar thermal devices in the correct position, both vertically and horizontally, designers will need to know about something called the **angle of incidence.**

In this landscape, look left to the house and then right to the tree; this is what is meant by horizontal orientation.

Next, look to the line demarcating the base of the hills from the plains far in the distance, then look up to the Sun; this is vertical orientation.

Angle of Incidence

This term sounds a bit scary but the concept that goes with it is really pretty easy to grasp, it is simply a measure in degrees of how far beyond perpendicular the Sun's ray are when they strike a collecting surface. The term perpendicular means that two lines or objects meet to form a 90–degree angle, also called a right angle. In solar thermal design it is when the Sun is shining straight at the collector's surface.

Though such terms as angle of incidence sound like big science lingo, they are useful to know. Say for instance that you are someone writing a book about solar designing. When you discuss things in which you need to refer to the angle of incidence between a collecting surface and the Sun, you can write the "angle of incidence" or you can write "how far beyond perpendicular the Sun's ray are when they strike the collecting surface" every time you need to refer to … the angle of incidence. I find that less words are usually better, so I'll stick with the term 'angle of incidence' throughout the book, unless I think a longer description helps.

Perpendicular alignment has an angle of incidence of zero and this is the position that will gather the most sunlight. The imaginary line that extends straight out from a surface in a perpendicular alignment is referred to as the **normal**. It is the amount of deviation from normal that the angle of incidence is a measure of.

If a collector could remain perpendicular to the Sun (in all of its planes) 100% of the time, the designer would find that they are in solar designing heaven, however, with a fixed device that situation can never happen. With an understanding that an angle of incidence of up to 10–degrees will lose them less than 2% of the total potential sunlight available and an angle of incidence of 60–degrees will lose them 50% of the potential available, the designer can make the right choice in regard to positioning the device. Knowledge allows them to make knowledgeable trade-offs. With a 50% loss at a 60–degree angle of incidence, the average designer won't even consider the placement unless it's either the only choice they have or they can design in some form of compensation – perhaps additional collecting area.

Good solar thermal designers need this basic information to make them good; it isn't luck.

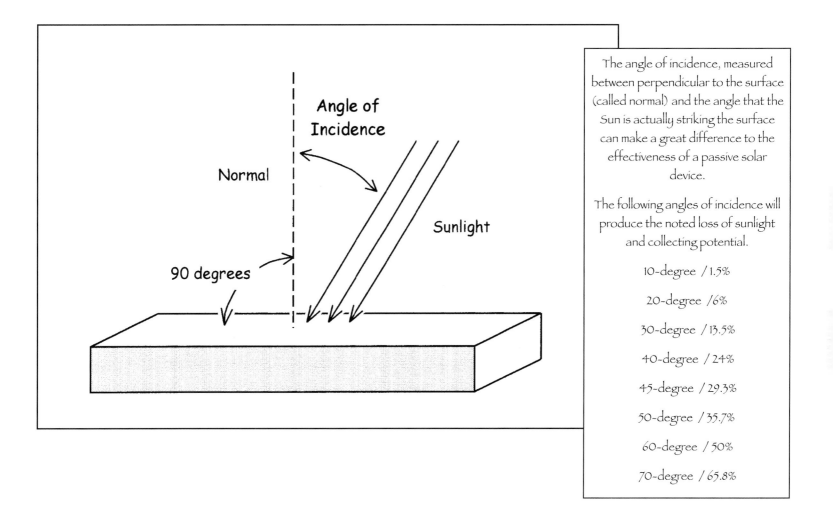

Angle of
Incidence

Normal

Sunlight

90 degrees

The angle of incidence, measured between perpendicular to the surface (called normal) and the angle that the Sun is actually striking the surface can make a great difference to the effectiveness of a passive solar device.

The following angles of incidence will produce the noted loss of sunlight and collecting potential.

10-degree / 1.5%

20-degree / 6%

30-degree / 13.5%

40-degree / 24%

45-degree / 29.3%

50-degree / 35.7%

60-degree / 50%

70-degree / 65.8%

East/West

Because the Sun passes overhead during the day, our potential ability to collect solar radiation changes every minute, all day long. Thankfully, Earth is rotating at a predictable speed and the Sun consistently moves through approximately 15 degrees of sky every hour. It actually moves through a bit more sometimes, a bit less at others, but we can safely rely on 15 degrees as the consistent average. With that information we can begin to understand the complexities of the Earth/Sun dynamic in relation to our goal of utilising the Sun as a heat source. As might be expected, the sunlight is strongest in the middle of the day (for reasons we'll get to in a few more pages) so the logical position in which to place a device with a fixed collector is in the direction of the midday Sun, which is also in direct alignment with the planetary pole for whichever hemisphere a device is in.

Thanks to the way national governments play with the duration of sunlight that is available at different times of the year (daylight savings time) a true midday Sun position is usually not at midday. The use of a compass and knowledge of the correct declination for the location will allow a true polar reckoning to be found. In spite of the fact that numerous young scouts of all ages manage this skill, I find that I am never really confident with my measurements. I need a more obvious way of determining the true midday Sun, something that I can see and something that is not influenced by the planet's magnetic fields. While trying to resolve this dilemma I did find what I was looking for. It is knowledge that goes way back; so far back in time that no one knows for sure the point of origin.

The method I learned to use for finding the location of the true midday Sun position for any place on Earth takes just a few hours of a single day. It just might be considered easier than using a compass with declination tables and in their place, this method involves the use of a clever little device called a **gnomon**. Gnomon is a 'big-city' term for a stick, but when used properly a stick can become a gnomon; it is a gnomon that can be found at the heart of a sundial, not a stick.

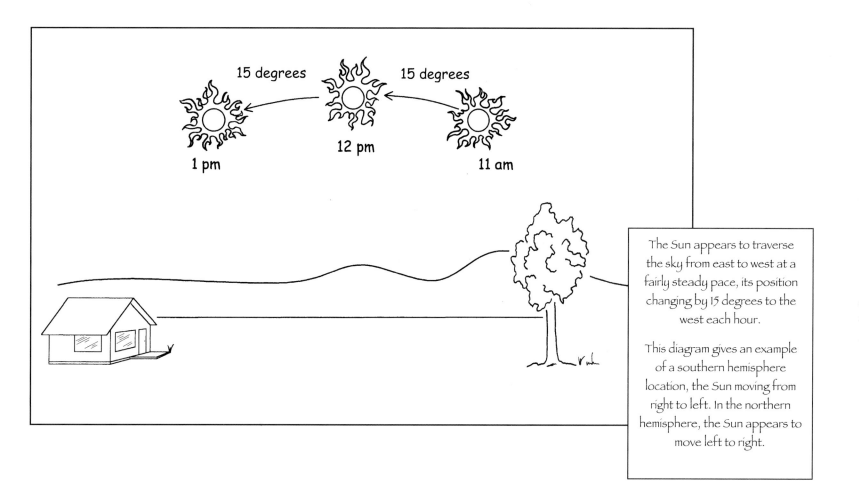

15 degrees   15 degrees

1 pm   12 pm   11 am

The Sun appears to traverse the sky from east to west at a fairly steady pace, its position changing by 15 degrees to the west each hour.

This diagram gives an example of a southern hemisphere location, the Sun moving from right to left. In the northern hemisphere, the Sun appears to move left to right.

Using a gnomon to locate the midday Sun position for a site begins by driving a suitably sturdy stick into the ground. Next, the stick must be confirmed to be in a perpendicular alignment to the ground and this can be achieved by checking it on several sides with a builders square or similar device that measures right angles. Easy. The next critical requirement is a sunny day and a good book.

Having had a good start with a sunny day, beginning in the late morning mark out on the ground where the tip of gnomon's shadow is every ten to fifteen minutes. Don't worry if there is no clock or timepiece to reference this interval with, just make the marks at whatever feels like that amount of time. This is where the good book comes in as sitting and watching a shadow move can be far from an exhilarating experience. Better to rely on the book to drive off near certain boredom as the Sun and the gnomon's shadow do their thing.

Skipping forward now to later in the day … some time ago a series of marks would have been made that are obviously closer to the gnomon than the others. Once the shadow has passed the position that resulted in these marks, the hard work is done. Carefully connect the marks to make a line that follows the path of gnomon tip's shadow. For all but two days of the year the shadow's line will form a curve. If this process is carried out on June 21st and December 21st, the lines will show the path of the Sun on the longest day (summer solstice) and the shortest day (winter solstice). On the summer side, the line will curve in a fashion that wraps around the gnomon; on the winter side the line will be farther from the gnomon and will curve away from it. The day that falls precisely between them, and the day with an equally long day and night period will create a straight line. This 'straight line day', which occurs twice a year, is known as the vernal (spring) and autumnal equinox.

To finish the job we'll need to draw an intersecting line from the base of the gnomon to the closest point on the shadow's line that we have drawn. That intersecting line is now running from the shadow's path, through the gnomon and pointing directly to the pole of whichever hemisphere the measurements are being made in. This direction is also where the true midday Sun position can be found. How easy is that?

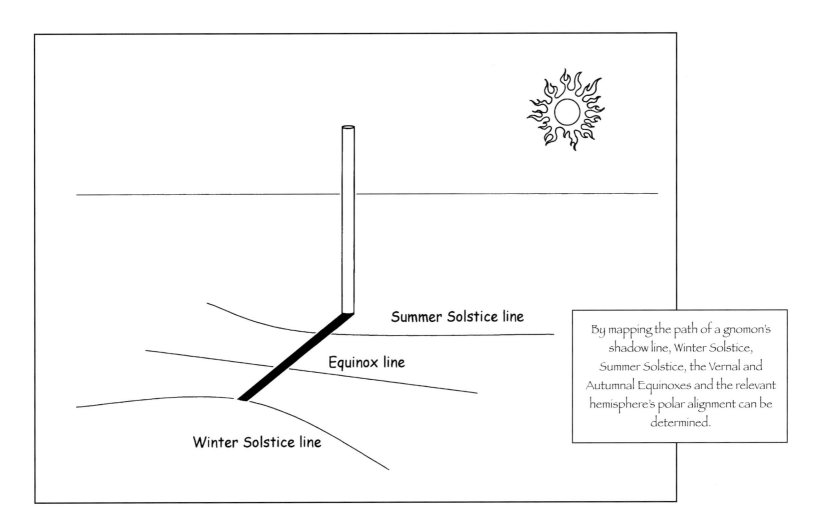

Summer Solstice line

Equinox line

Winter Solstice line

By mapping the path of a gnomon's shadow line, Winter Solstice, Summer Solstice, the Vernal and Autumnal Equinoxes and the relevant hemisphere's polar alignment can be determined.

Longitude

The Earth is a ball; a beautiful blue sphere spinning around in spacetime, orbiting the Sun in an elliptical path once every year and residing in a happy solar system of nine known planets that are also orbiting the Sun in typically elliptical orbits. Planetary orbits range in duration from that of Mercury, the closest planet to the Sun, making the round trip in just 176 Earth days to Pluto, the smallest of all confirmed planets and sometimes-farthest from the Sun, taking 248 Earth years to make a full orbit. Pluto is said to be the 'sometimes farthest' because it is orbiting in such an extended ellipse that during its 248 Earth-year orbit, the small planet occasionally swaps places with Neptune, typically in the number eight position, as the planet farthest from the Sun. Pluto's orbital path brings it inside of Neptune's orbit and the two change the status of the farthest-out position for about twenty years before Pluto's orbit sweeps back outside of Neptune's, regaining Pluto the farthest-out position once again.

In response to the effects of gravity in a weightless environment, the Sun, all nine of our solar systems planets and probably everything else in the universe that formed from either, gases, liquids or fields of floating debris are round and rotate on an axis, as if a stick is driven through them just before they were set to spinning. If our planet was a tomato, used as an example because a tomato is also round (well, kind of round) we could consider that the rotational axis ran through the tomato from the stem to the butt. Let's say the tomato's stem is the north pole and its butt is the south pole of an imaginary rotational axis.

Supposing that someone was making a salad for us and we wanted to have the tomato sliced into wedges with the cuts running from the stem to the butt, lines could be marked on its surface to show that person exactly where to make the cuts. Lines in this orientation on the tomato would be in the same position as lines of **longitude** that have been drawn by mapmakers on their renditions of the Earth's surface. The Earth's lines of longitude run from the North Pole to the South Pole and dissect the planet into a series of wedges.

With people needing to have references such as this for navigation, mapmakers have given lines of longitude permanent designations, starting from a single line that has been assigned zero degrees as its location on the surface of planet Earth. This line is known as the **prime meridian**.

Lines of longitude are also called **meridian lines**, or simply meridians, and each line travels halfway around the Earth – pole to pole. Back in years gone past, where to start counting was quite a confusing issue; there is no natural starting or stopping point for east and west. Sailors and mapmakers once referred to a longitude line drawn through the city or town that they were from as *their* zero meridian. To end the confusion, in 1884 the International Conference in Washington imposed upon the entire world that the line running through Greenwich, England would be zero, the Prime Meridian and the line from which all others would be measured. Being quite a sensible lot, sailors and mapmakers had already chosen to split the planetary sphere up into 360 longitude lines, one for each of the 360 degrees of a circle. On a finer scale, lines relating to portions of degrees, called minutes and seconds, are also noted.

From the Prime Meridian, travellers going west enter into the Western Hemisphere and can go halfway around the Earth to the 180 degrees west longitude line. Travellers heading east from the prime meridian can go halfway around the Earth to the 180 degrees east longitude line. Where east meets west …the 180 degree line … it's the same line.

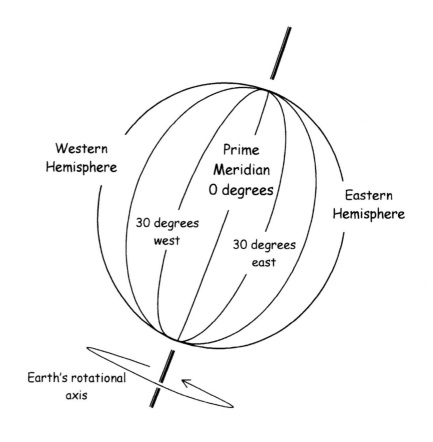

Longitude, Gnomons and Solar Alignment

Back at the site where the gnomon was used to determine true north, the midday reference line is useful to solar designers because when the Sun is in alignment with a location's true polar orientation, it is also in the middle of the sky. It will rise no higher on that day and in this position the Sun is also directly aligned above the site's line of longitude. Its rays are strongest at a given site when in this alignment because the Sun is as close to perpendicular with the ground as it will be and therefore providing the lowest angle of incidence against the sphere of Earth. The rays of light are also coming through the least amount of atmosphere in this final part of their journey from the Sun to us. Before making contact with the planet's surface the rays of light must pass through our atmosphere and when they are coming in at a perpendicular alignment to a given point on the surface, there is less obstruction by the gasses, water vapor, dust and other elements that are floating around up in our sky.

While a midday Sun orientation is a key factor to consider, other criterion that will determine the final positioning of a collecting surface are largely localised. Wherever there are hills or trees that might be an obstruction or typical cloud and fog patterns at a certain time of the day, the orientation should be adjusted east or west. This tweaking of the device's position will ensure that the least angle of incidence in the horizontal plane occurs between the collecting surface and the Sun's rays during a more favorable point in the day; either earlier or later than midday.

If the goal for the installation is to achieve 90% or better collecting ability for the total available sunlight during a particular period of the year – say mid-winter – then a true polar alignment is the best position in which to place our devices. Reality isn't always so kind to such ideals but it is comforting to know that devices can be up to 15 degrees east or west of the true midday Sun position with only a couple percent of additional loss resulting from the greater atmospheric thickness. Such compromises are just part of the process involved when working with Nature in order to determine the best horizontal alignment for devices that collect solar heat.

Summer/Winter Considerations

To make the most of what both summer and winter have to offer in the way of available sunlight, designers need a few more pieces of information so they can make informed decisions in regard to a device's vertical orientation. Great mathematical calculations or scientific qualifications are not required in order to determine a suitable angle in which to fix a solar collecting surface's vertical orientation and it would be unfair to say that solar heated devices would not work well in the winter months; as my intro story stands as testament, even in the permanently frozen state of Antarctica there is plenty of potential for collecting and using the Sun's warmth. Because winter is usually the time when warmth is hardest to come by, it makes the most sense in most situations to optimise a fixed collector device for use during this period.

The key to maximum collecting in the coldest months of winter hinges upon the designer's ability to calculate a few angles, based on the Sun's position in the sky. As has already been mentioned, the Sun will have a low position in relation to the horizon at the winter solstice and it will be high overhead during the summer solstice. This occurs everywhere on the entire planet – even at the poles. Information is once again the key to success and in this case, success can be found in a situation that is close to ideal.

Remembering that as long as a collector's angle of incidence varies by no more than 10 degrees from an optimum setting, the loss of potential collecting ability is only about 2%, a solar designer working on a fixed collector device will play the averages to create a 'best possible' orientation. Losses on a magnitude of just two percent are pretty reasonable to accept when working with solar heating; realistically they are not even noticeable and honestly, most installations can cope with such a minor loss. So now, tapping into an age-old saying, *"close only counts in horseshoes and hand grenades"*, we can also add … *"and solar thermal design"*.

The simple way to set the collecting surface's angle close enough, continues with the determination of the device's **latitude**.

Latitude

Consider for a moment that the planet, our round sphere known as Earth, is a rather round person with their head at the North Pole and their feet at the South Pole. If we were to draw a circle around the sphere of the Earth exactly between its head and feet, lets say 'the Earth's belly', the line marking that circle would be what is called the equator. The equator forms a line, which is the dividing line that separates the northern and southern hemispheres. The equator and all other lines running in this general direction are called lines of latitude and have been assigned a designated position based on a system of degrees, like the lines of longitude.

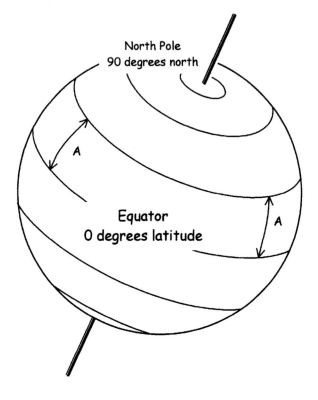

The equator has been assigned a latitude designation of zero degrees. Each of the poles has been assigned a latitude designation of 90 degrees; 90 degrees north and 90 degrees south. For anywhere else on the planet, latitude is measured in degrees based on the circumference of the Earth, like a circle drawn from pole to pole with its base point at the very center of the planet. Between the equator and the pole of whatever hemisphere a person happens to be in, latitude will be somewhere between 0 and 90 degrees and as with longitude, latitude also uses minute and second lines, for locations spotting on a finer scale.

Because lines of latitude run around the globe in an east/west orientation, they are parallel to the equator and always maintain an equal distance from the equator (as well as any other particular latitude line) as shown by the measurement 'A' in the diagram above. Lines of latitude have therefore also been called **parallels**. Parallel lines running in the same direction can never meet so each is unique in having its own slice of the planet, unlike the longitudes, which meet at the poles.

Although the degree of latitude cannot be determined for a particular site by natural events as easily as when working with longitude, the hard work of determining latitude was done long ago. Nowadays it is very easy to find the latitude of a site – just look on a map. Determining the optimum angle for a fixed, solar collector's **vertical orientation** is also … not too hard. If it has been determined that the device needs to operate at its best year round then, as mentioned, mid-winter is the likely time to focus on for optimisation. Actually, in most places the coldest period of the year is the two months occurring after mid-winter but this varies of course, so local considerations have to be taken into account. A personal knowledge of what kind of ambient air temperatures to expect in relation to the months around the mid-winter period for a chosen location is vital.

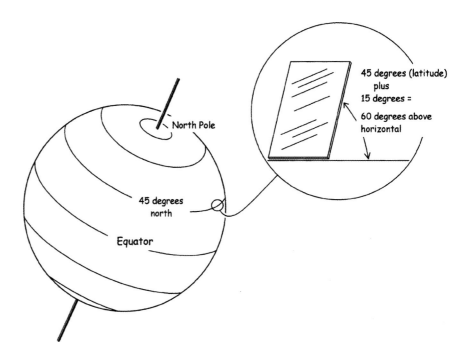

With a map in hand, search out the location of the site where the device will be used and find its latitude. Assuming that mid-winter is the chosen optimisation period, as in the diagram to the right, add 15 degrees to the site's latitude to determine the necessary angle for the collector to be positioned above horizontal or lying flat on the ground. Why? That's easy to explain …

On Midwinter's Day, the Sun's path will have actually dipped low enough that the Sun will be below a perpendicular alignment with the collecting surface at midday. It will stay below perpendicular for a couple more weeks before beginning to rise back into that more favorable alignment. By using this calculation to determine vertical orientation, a solar collecting device will achieve maximum collecting potential for the six to eight weeks beginning about a month after Midwinter's Day, just as that coldest part of winter usually sets in.

Bringing it all Together

With all of this learning behind us now – planetary axis tilt, elliptical orbits, angle of incidence losses, gnomons, longitude, latitude – we can reason that in relation to the Sun's path (which moves through 15 degrees of sky per hour) the best solar exposure for a given site will only occur for about three hours each day; from a time one and a half hours before true midday to a one and a half hours after true midday. The duration for this period of optimum daily potential will be the same regardless of whether it is summer or winter although in the summer, with the Sun high overhead and coming through less of Earth's atmosphere, there may be more intense sunlight available and therefore an ability to generate higher temperatures.

When working with a fixed collector's horizontal orientation, good solar designers will realise that there is only this three-hour window of opportunity each day to collect the most sunlight possible at a given site. Some designers refer to this as a 'sweet spot' in the sky. By relying on polar alignment for positioning the device's horizontal orientation, the best designers will consider the most favorable ways to use the sweet spot when making this decision.

Calculations for a fixed collector, positioned no farther than 15 degrees east or west of a true polar alignment *and* optimised for winter season exposure by setting its vertical orientation at an angle 15 degrees higher than the latitude of the site, indicate that the device can collect the maximum solar energy available for approximately 252 hours each year. That breaks down to three hours a day, seven days in a row, for twelve weeks in a row, just after the middle of winter. Twelve weeks is one-quarter of the year so a fixed-collector device, optimised for mid-winter, is capable of gathering about a quarter of the maximum available solar energy available to it year round. In this position it will experience a shortcoming in ability during the middle of summer, when the Sun is high overhead.

Should we be content to waste collecting potential in the middle of summer, at what is likely to be the best time of the year for available hours of sunlight? The term 'waste' is probably not too harsh. Consider that if the device were fixed in the horizontal plane, but adjusted daily for the best orientation in the vertical plane; it would then have approximately four times as much collecting ability than if it were fully fixed. And what if the device was able to track the Sun from east to west, through the sky each day … well, that difference is not very comfortable to think about if our device is unable to do it.

As mentioned a couple pages back, when we get the *Practical Applications* section later in the book, I have some details on a device that can allow tracking of the Sun in both the horizontal and vertical planes. We still must consider though, that regardless of the planet's seasonal position and no matter what time of day it is, once we've done the calculations to orient a device as well as possible, the ultimate performance of the device is then out of our hands. There's always that 'one more thing' to think about – just to make life interesting and as if we didn't have enough considerations going around in our minds already. As we know conditions change regularly here on planet Earth and in spite of our best efforts, we still need to hope that it isn't cloudy when we want to be collecting.

Just for now, we have a little bit of history coming up …

Chapter Two

A History of Solar Powered Invention

Some of the first records of solar design come from the ancient Greeks. It is recorded that the great thinker Socrates (469–399 B.C.) designed a house that maximized solar exposure in the wintertime to help keep it warm. By fixing windows on the sunny, south facing side of the house and creating a heavy wall with insulating properties on the north side, the house would collect warmth from the Sun and hold it well. 400 years before the birth of Christ, Socrates designed the world's first recorded passive solar dwelling.

In 300 B.C., the Greek mathematician Euclid explored the reflective abilities of curved surfaces and Archimedes, a Greek philosopher who lived from 287–212 B.C., carried this work to new heights. Archimedes used curved reflective panels to concentrate the Sun's light and set fire to the invading Roman ships in a battle at Syracuse, Sicily in the year 214 B.C. Some Centuries later Leonardo da Vinci suggested that sunlight be used as a heat source for commercial purposes.

With such successes having been recorded well before the beginning of our modern time line (A.D.) and continuing support for the industrial use of the Sun coming from people like da Vinci, one might think that the world was on its way to energy independence hundreds of years ago. Alas, technology and the acceptance of technology by the very people that it can benefit, has a way of taking its own time. As the centuries rolled past, it seems that scientists of nearly every great period of discovery since the Dark Ages shouted "Eureka!" as they 'discovered' solar power, but somehow the knowledge managed to remain an obscure notion to the masses of common people whose lives would have benefited by its adoption for things like home heating, cooking and leading a generally happy existence.

Archimedes

Although he is considered to be one of the greatest mathematicians of all time, it could be said that Archimedes was also an ultimate do-it-yourselfer. His life's accomplishments are remarkable and range from the practical to the mathematically complex. He did much of his work by drawing pictures and diagrams to assist his reckonings and was committed to this method to the point of carrying around a pan full of sand, just in case an inspiration came to him unexpectedly.

Accomplishments:

- Defined pi by drawing and inscribing a circle with regular polygons that had 96 sides;

- Invented the Archimedes Screw, a fluid pump that consists of a helix fixed to a shaft and revolving inside of a fixed cylinder or tube. The device is still in use, especially used for pumping sewage in wastewater treatment plants;

- Discovered Law of Hydrostatics (buoyancy), also known as the Archimedes Principle. Basically, any object immersed in fluid, is buoyed up by a force equal to the weight of the fluid displaced by the object;

- Developed the compound pulley, precursor to block and tackle pulleys;

- Invented the catapult;

- Discovered fundamental theorems concerning the center of gravity of plane figures and solids;

- Designed the solar concentrating war mirrors of Syracuse – possibly the first use of solar power other than for general warming.

It is sad to note that two years after the successful defence of Syracuse, largely through the use of the burning mirrors, Archimedes was killed in his chambers when the Romans did finally succeed in taking the city. Although the exact details of his death are still speculated, it is largely accepted that he was killed by a Roman soldier. Apparently he refused to leave with the intruder as he was busy working out some sort of mathematical problem. Outraged by the refusal, the soldier ran him through with his sword.

The great thinker and mathematician, Archimedes. He was possibly the first person to actually harness the power of sunlight for human purposes other than staying warm.

Horace de Saussure

In the mid 1700's, Horace de Saussure (1740–1799) a French-Swiss naturalist, wrote of his observations that a room, a carriage, or any other place enclosed by glass becomes hotter when sunlight comes into the space through the glass. While scientists of the time worked with sunlight and mirrors to perform such then-amazing feats as burning objects at a distance, boiling water to make steam or melting metals, de Saussure was surprised that the heating phenomenon noted in glass enclosures had not led to any notable research projects. In 1767 de Saussure set out to determine how effective glass-covered heat traps were at collecting the energy in sunlight and to investigate the maximum temperature attainable in such a device.

His first experiments were conducted with the use of a miniature greenhouse that he constructed. The model consisted of five square, glass boxes with an overall size of 12-inches by 6-inches (300mm x 150mm), with the innermost box being 4-inches by 2-inches (100mm x 50mm). For his experiments, the boxes were stacked one inside the other, atop a black wooden table and after exposing the apparatus to the Sun for several hours, de Saussure would measure the temperature inside of each box. As we might now expect, he found the outermost box to be the coolest, with the temperature increasing inside each succeeding box.

An interesting aspect of this experiment was that de Saussure would rotate the assembly of boxes so that solar rays always struck the glass covers of the boxes in a perpendicular plane to allow them to gain maximum heating effect. This was a very early use, perhaps even the first recorded use, of a manual solar tracking system. The highest recorded temperature in the experiments was 189.5°F (87.5°C) and eventually he used the experimental apparatus to cook food, notably fruit.

Physicists of de Saussure's time were still unsure of the nature of sunlight. With the study of radiation and the physical properties of light still centuries away, science couldn't offer him any explanations as to how the light made heat. Though the warming effect of sunlight could be felt and his experiments demonstrated that the heat could be captured, de Saussure remained uncertain of how the Sun actually heated the air inside of the glass boxes. In spite of the lack of theoretical understanding, the validity of the results provided by these tests stand for themselves; with the wisdom of hindsight, we can now say that in 1767 Horace de Saussure invented the first flat plate solar collector.

Horace de Saussure conducted the first recorded experiments to determine how and why sunlight created warmth in an enclosed space.

Antoine LaVoisier

Antoine LaVoisier (1743–1794) was a French chemist that became known as the father of modern chemistry. He was famous for his experiments in combustion of materials and his range of discoveries is vast. While in his early twenties he married the 13-year-old daughter of one of his workmates, a controversial situation with a girl named Marie-Anne. Proving their relationship to be more than a youthful folly, throughout their life together she worked with him as his laboratory assistant, translated texts from English for him and illustrated his books.

Through an experiment in burning phosphorus and sulphur in air, he proved that the resulting products weighed more than the original ingredients. Further results from this experiment's findings revealed that while his burnt products gained mass, the air in which they were burnt lost mass, forming the basis of the scientific Law of Conservation of Mass.

In other experiments he demonstrated that air, which is necessary for combustion to take place, consisted of several elements; one of which he named oxygen. We also still refer to the name he gave to the 'explosive air' discovered by his contemporary researcher, the Englishman Henry Cavendish – hydrogen. He invented the system of chemical nomenclature that is still in use today, including names such as sulphuric, sulphates, and sulphites and his 1789 book, *Traité Élémentaire de Chimie* was the first modern chemical textbook. It contained a clear statement of the Law of Conservation of Mass and included the first listing of elements or substances that could not be broken down further including oxygen, nitrogen, hydrogen, phosphorus, mercury, zinc, and sulphur.

Why am I going on about him in this book? In his trials of ways to burn things, LaVoisier invented a solar furnace that used two huge focusing lenses that caught the Sun's rays, amplified their strength and sent them into the furnace. Temperatures within LaVoisier's solar furnace reached between 2,000 and 3,000 degrees Fahrenheit, suitable for melting a range of metals. Some texts indicate that his solar furnace was able to melt platinum, which has a melting point of 3,236°F (1780°C), providing us with a very early account of a person manipulating sunlight to reach the temperature levels necessary for industrial use.

Sadly for the world of science, he was also a French tax collector during the era of the Revolution and was beheaded as an enemy of the people.

Antoine LaVoisier. His early experiments in combustion of materials lead to the development of a solar furnace.

Sir John Herschel and Samuel Pierpont Langley

Skipping into the next century, two notable nineteenth-century scientists conducted experiments with hot boxes similar to that of Horace de Saussure. In the 1830's the noted astronomer Sir John Herschel (1792–1871) made a hot box while on an expedition to the Cape of Good Hope in South Africa. Constructed from mahogany timber and blackened on the inside, the box was covered with glass, set into a wooden frame that was protected by another sheet of glass and placed on the ground where sand was heaped up along its sides. The outcome of his experiments was not only scientifically interesting but also, like de Saussure's final trials, edible. His notes indicate that temperatures within the box rose to 240°F (116°C) and members in his group found some level of amusement in placing eggs, meat, and other foods inside the Sun-heated chamber. To their delight, all foods were found (after a reasonable period of time) to be cooked to their satisfaction.

Samuel Pierpont Langley (1831–1906), an American astrophysicist who later became head of the Smithsonian Institution, had been interested by solar-generated heat since he was young. Wondering why glass coverings kept the interior of a greenhouse warm, in 1881 Langley took a trip to the highest point in the continental United States, California's Mt. Whitney (elevation 14,491 feet/4416 metres). He chose the mountain so he could study the effects of sunlight's warming ability in a range of temperature environments. The experiments were conducted with a hot box and Langley described his results and experiences, which were also very similar to Herschel and de Saussure's, in an 1882 issue of the periodical, *Nature*.

Langley's narrative noted that as his group climbed the mountain and the surface temperature of the soil fell to the freezing point, the temperature within a copper vessel over which he had placed two sheets of ordinary window glass, rose above the boiling point of water. Eureka? Langley discovered that he could boil water by the power in the Sun's rays; actually more of a validation of the discoveries that had been made almost a century earlier by de Saussure and LaVoisier.

Though advances in solar powered devices during the preceding centuries was slow to non-existent, the 1800's basically saw old ideas revisited, new ones developed and then all kinds of ideas continually re-examined without much occurring in the way of practical application. The coming pages will look at a few of the highpoints of solar powered device development within this era as well as some significant implementations of devices in the early 1900's.

Sir John Herschel. Among many
other scientific achievements credited
to him, he also experimented with
solar heating.

A History of Solar Powered Invention

August Bernard Mouchot

In 1864, 17 years before Langley's trip to Mt. Whitney, the Frenchman August Bernard Mouchot, a mathematics instructor, devised a water pump that ran from solar energy collected in a 16 ½ foot (5 metre) wide, reflective dish. Expressing his concerns about France's dependence on coal, Mouchot wrote: " ... *eventually industry will no longer find in Europe the resources to satisfy its prodigious expansion. Coal will undoubtedly be used up. What will industry do then?*"

In his early experiments, Mouchot attempted to gather the Sun's heat in a glass-encased, iron cauldron. Sunlight would pass through the glass, heating the iron cauldron and subsequently the water contained within. Mouchot's device did bring water to a boil but the quantities and pressures of steam produced by the system were not enough to do productive work, such as operating a steam engine. Through further experimentation he discovered that by adding a cone-shaped, concentrating reflector to the assembly he could raise the temperature of the cauldron and generate more steam. His reflector was unique from spherical concentrators that focus light to a single point. Mouchot's 'truncated cone' design focused light along the full length of the centrally mounted, cylindrical boiler. This arrangement resulted in a lower but more controllable heat than that generated at a single focal point. This innovation successfully improved the machine, which pumped over 2 tons of water per minute.

In late 1865, he succeeded in using his apparatus to operate a small, conventional steam engine, pleasing Emperor Napoleon III to the point that the monarch offered financial assistance to develop an industrial version of the solar motor. With a range of improvements added – increased capacity, a refined reflector and a solar tracking mechanism – in 1872 Mouchot's device produced a one-half horsepower output while operating a steam driven water pump in the library courtyard of his home.

For government-backed agricultural trials of the machine in the French protectorate of Algeria, a sunny region that was entirely dependent on coal for industrial power production, Mouchot decided to enlarge its capacity to 22 gallons (100 litres) and to use a multi-tubed boiler that would produce more pressure and improve the steam engine's performance. While the success of the device was acknowledged, during the time that it was being redeveloped and the Algerian installation completed, the price of coal and its transport had dropped for France. With the need for alternative fuel sources becoming a lower priority, in 1881 the French government deemed the device was not worth pursuing any further and Mouchot returned to his other academic interests.

August Mouchot and his truncated cone reflector. Mouchot's reflector/boiler combination was the heart of the first device recorded that could generate enough steam pressure to operate engines.

A History of Solar Powered Invention

William Adams

William Adams was the Deputy Registrar for the English Crown in Bombay, India, during the 1870's. With an interest in the solar power experiments of the time, he recognised several difficulties in constructing a large, cone-shaped, polished metal reflector like Mouchot's. He felt that in order for it to be large enough to generate more than one-half horsepower, the reflector would tarnish easily, be too costly to build and difficult to manoeuvre – but he had his own ideas.

Adams was convinced that a reflector constructed of flat, silver-coated mirrors arranged in a semicircle around the boiler would be cheaper to construct and easier to work with. His plan was to build a rack that would hold the many small, flat mirrors that would reflect sunlight in the direction of a stationary boiler. During the designing process, Adams calculated that the reflector for each boiler would require 72, 17-inch by 10-inch (430mm x 250mm) flat mirrors to generate the level of heat required for industrial-grade steam pressures (approximately 1,200°F / 650°C). This design retained Mouchot's glass-enclosed cauldron but removed the weight of the boiler unit from the moveable part of the machine. To track the Sun's movement, Adams would employ a laborer to roll the rack of mirrors around a semicircular track in order to keep the Sun's rays shining on the boiler.

Construction began in late 1878 and as planned, he arranged the collecting mirrors around a stationary boiler that was connected to a 2 ½ horsepower steam engine. The boiler produced industrial grade steam and his machine successfully operated for a two-week period. With this experimental success, Adams was eager to display his invention. He notified newspapers and invited his important military and government friends to a demonstration. All were apparently impressed; even those who, while doubtful that solar power could compete directly with coal and wood, agreed that use of the Sun could be practical as a supplemental source of power.

For unknown reasons, Adams' experimentation ended soon after the demonstration, leaving us with a rather hollow legacy to a remarkable design and testing process. His efforts to create a practical and useful method of gathering solar heat are not lost however. He wrote an award-winning book entitled *Solar Heat - A Substitute for Fuel in Tropical Countries* and modern engineers refer to Adams' design as the Power Tower concept. It is recognised as one of the best configurations for large-scale, centralized solar plants that generate steam to drive an accompanying engine.

Power Tower System

Many individual reflectors are placed around a tower and are continually adjusted by a solar tracking system so that they reflect sunlight at the central receiver. The receiver then heats either water or molten nitrate salt to power a steam turbine generator.

This power tower system had been installed in New Mexico as a municipal power generation facility. It is no longer in service.

Photo courtesy of National Renewable Energy Lab
Image credit: Sandia National Labs

John Ericsson

Though Swedish by birth, the American inventor John Ericsson (1803–1889), whose most celebrated accomplishment was the Civil War battleship "Monitor", dedicated the latter part of his career to peaceful pursuits such as solar power. Ericsson shared the growing fear that someday coal supplies would run out. In 1868 he wrote of his belief that within a couple of thousand years the coal fields of Europe would be completely exhausted unless the Sun's heat was used wherever practical.

In 1870 Ericsson developed what he claimed to be the first solar-powered steam engine, dismissing Mouchot's machine as "a mere toy". Funnily enough, Ericsson's first designs employed a conical, dish-shaped reflector that concentrated solar radiation onto a mounted boiler and utilised a tracking mechanism that kept the reflector directed toward the Sun; very 'Mouchot' indeed.

Ericsson did come to invent a new method of collecting solar rays however – the parabolic trough. Unlike a true parabola, which is dish shaped and focuses a broad field of light to a single, relatively small point, a parabolic trough is more like a tube cut in half lengthwise. The trough has a focus that runs along a line in front of the open side of the half-tube. Though the trough's temperatures and efficiencies were not as high as that experienced with dish-shaped reflectors, it offered many advantages over its dish-shaped counterparts. The trough was simple to design, inexpensive to construct and only needed to track the Sun in an up and down motion if lying horizontal or east to west if mounted vertically, making the mechanisms inexpensive to construct.

By 1870 he had constructed a linear boiler (essentially a pipe) and placed it in the focus line of the trough. In trials he positioned the new arrangement toward the Sun and, like his contemporaries, connected it to a conventional steam engine. The machine ran successfully, though he declined to provide power ratings at the time. Later studies of Ercisson's system believe that it produced about 2 ½ horsepower – indeed a step up from the installations of both Mouchot and Adams.

Ericsson worked for many years on a design for a "sunpower station" that never eventuated but his new collection system became popular with later experimenters and engineers. It became the core of the world's first large scale, Sun-powered application, constructed by Frank Shuman in 1906, and is as much a standard for modern plants as Adams' Power Tower design.

Parabolic troughs operating in the Mojave Desert are a modern application of the design pioneered by John Ericsson (above).

A History of Solar Powered Invention

Charles Tellier

With Mouchot's field application work having been halted by the French government, Adams experiments in India slowly fading into obscurity and Ericsson's work breaking new ground but basically creating new adaptations from old ideas, the designing of a new kind of device for solar heat collection fell to the notable French engineer, Charles Tellier (1828–1913). Considered to be the prime inventor of refrigeration systems, Tellier also designed the first non-concentrating, non-reflecting, solar powered device, which operated from a collecting unit that was very similar to modern flat plate collectors.

In 1885, Tellier installed a solar collector on his roof that was composed of ten small, metal containers that were plumbed to one another to form the collecting unit. Each container consisted of two iron sheets that were riveted together to form a watertight seal and fabricated with the necessary plumbing fittings to accommodate the connecting pipes. Tellier made a radical departure from previous work by using ammonia as the heated fluid in his system rather than water. Ammonia has a significantly lower boiling point than water and after exposure to the Sun his collectors emitted enough pressurized ammonia gas to operate a pump that moved water at the rate of some 300 gallons (1350 litres) per hour. Tellier thought that by simply adding more collecting plates to increase the size of the system, he would finally make affordable, solar powered industrial pumping applications possible. By the end of the 1880's he had improved the efficiency of his system by enclosing the collecting plate assembly within a glass-covered casing that was insulated on the bottom. He published his test results, which included details on his intentions to use solar powered systems to manufacture ice.

Tellier also believed that in spite of reductions in the price of coal to France, a consistent and readily available supply of energy was still needed for the French territories in Africa. Looking for larger applications for his invention, he proposed that the construction costs of his system, centered around the flat plate collector with ammonia-gas powered motors, were low enough to supersede Mouchot's invention and justify its implementation in the African territories. Apparently the French government thought differently and Tellier decided to continue with his refrigeration interests rather than carry on with solar experimenting. Once again we are left with a story where we can only wonder where the inventor's solar designs may have ended after such a promising start.

15 F
REPUBLIQUE FRANÇAISE
POSTES
LE FRIGORIFIQUE
1828
1913
CHARLES TELLIER — INDUSTRIE DU FROID

Though he was a key inventor in the development of solar powered devices, Charles Tellier is better known for his ground-breaking work in refrigeration.

Tellier transported a meat shipment from Buenos Aires to Rouen, France, aboard his own refrigerated steamer , Le Frigorifique, in 1876. The event is here commemorated on a French stamp in his honor.

Aubrey Eneas

Boston resident Aubrey Eneas had begun his solar experimenting in 1892. One of his first experimental efforts was based on the parabolic trough, but Eneas could not achieve high enough temperatures for industrial applications. Eventually he gave up on the troughs and turned his interest to cone-shaped reflectors like those used by Mouchot. While this approach resulted in higher temperatures, Eneas was still dissatisfied with the performance but decided to continue with the arrangement by redesigning the dish in order to achieve the performance temperatures that he desired. By altering the shape of the reflecting cone's sides to be more upright he was able to focus more sunlight on the boiler and acceptable performance was finally in his grasp.

He promoted the machine by exhibiting it in at Edwin Cawston's ostrich farm, a popular tourist attraction, in Pasadena, California. With a reflector that spanned 33 feet (10 metres) in diameter, contained 1,788 individual mirrors and a boiler that was about 13 feet (4 metres) long and held 100 gallons (455 litres) of water, the monstrous machine was a certain attention getter. Steam from the boiler operated an engine that pumped 1,400 gallons (6350 litres) of water per minute from a well on the farm and Eneas' marketing effort paid off as thousands of visitors left the farm convinced that the machine would soon be a fixture in the sunny Southwestern United States. Newspapers and science journals sent their reporters to the farm to cover the spectacle and Frank Millard, a reporter for the magazine *World's Work* predicted that the potential for this novel machine was not limited to agricultural irrigation. Millard suggested applications for solar powered steam as diverse as grinding grain, sawing lumber and running electric cars.

Eneas formed The Solar Motor Company in 1900. Begun in Boston, Massachusetts. He relocated to Los Angeles in 1903 so the company would be closer to potential customers in the southwestern region. Early the following year he sold his first complete system for $2,160.00 to Dr. A. J. Chandler of Mesa, Arizona. Unfortunately, after less than a week, the unit was damaged in a windstorm and collapsed. The boiler fell into the reflector and the machine was declared beyond repair. In the autumn of 1904, John May, a rancher in Wilcox, Arizona, bought one of Eneas' machines. Again, not long after purchase the machine was destroyed by bad weather. This second weather-related incident all but proved the beliefs of William Adams from nearly 30 years earlier in regard to the impractical nature of employing large parabolic reflectors. Unable to survive on the limited sales potential of a machine that couldn't withstand wind, in 1905 the company closed its operations.

Henry E. Willsie

Henry E. Willsie began his work with solar powered motors around the turn of the century. In his opinion, the failure of industry to accept the devices by Mouchot, Adams, Ericsson, and (later) Eneas proved that high-temperature, sunlight-concentrating machines were not practical options to pursue. He was convinced that a non-reflecting, lower-temperature collection system similar to Tellier's was the best method to use when capturing solar heat. Willsie also felt that solar motors would never be practical unless they could operate around the clock. Improvements in thermal storage techniques therefore became a hallmark of his experimentation.

Using sulphur dioxide as the active heat transfer medium in his system, Willsie built large, flat-plate collectors that first heated hundreds of gallons of water, which he kept in an insulated containment system. Once the water was up to the necessary operating temperature, a series of vaporizing pipes containing the sulphur dioxide were lowered into the basin. Heat passed through the pipes, transformed the sulphur dioxide into a high-pressure vapor, which after operating the engine passed to an exhaust system and finally into a condensing tube. The condenser cooled the gas and returned it to a liquid state, making it available for re-use.

In 1904, confident that his design would produce an uninterrupted power supply, he built two operating plants; the larger being a 15-horsepower operation in Needles, California. After several trials, Willsie was prepared to test the heat storage capacity of the system. After darkness had fallen, he opened a valve that allowed the solar-heated water to flow over the exchanger pipes and heat the sulphur dioxide contained within – the engine started. Willsie had created the first solar-heated device that could operate at night. His 15-horsepower machine was the most powerful system constructed up to that time and provided the unique feature (for a solar powered system) of being capable of continuous production; Willsie offered his machines for sale.

Unfortunately, there were no waiting buyers. Despite the favorable long-term cost analysis that revealed an expected two-year payback (which rivals even modern alternative energy systems) potential customers raised a number of important concerns: long-term durability, the machine's size in relation to power output, initial investment cost and others. With no buyers, his company closed without ever making a single sale.

Frank Shuman

Frank Shuman (1862–1917) produced his first effort at solar powered design in 1906. Like Willsie's device, Shuman's was also centered on a flat-plate collector and also breaking from the work of previous researchers, he used ether as the working fluid. Shuman's first machine performed poorly because even at industrial temperatures and pressures, the vapor that was produced from heated ether was not dense enough to drive a motor. Shuman's persistent nature would not let this initial failure keep him down in his search for a practical, solar powered application suitable for industrial use.

Shuman came to realize that he would need to produce steam, but he also believed that using complicated reflectors and tracking devices would be too costly and impractical for common use. Whether he took a leaf from Willsie's book or not, his similar answer was to conserve the heat already being gathered. Shuman enclosed his collector plates with dual panes of glass separated by a one-inch air space and replaced the boiler pipes with a thin, flat metal container similar to Tellier's original design. By 1910 the apparatus could consistently boil water but still was not capable of generating adequate pressure to drive an industrial-sized steam engine.

Faced with a need to capture still more heat, he gave in to using a reflector. To keep costs down he arranged two rows of ordinary, flat mirrors in order to more than double the amount of sunlight that was collected. In 1911 he constructed the largest solar-heated system to date, near his home in Tacony, Pennsylvania. With more than 10,000 square feet (6050 square metres) of collecting area, trials of the machine still did not provide high enough pressures. Not easily defeated, Shuman and engineer E.P. Haines redesigned the common industrial steam engine to operate at lower pressures. Their new engine design developed 33 horsepower from the system and drove a water pump that delivered 3,000 gallons (13,650 litres) per minute.

Like the French entrepreneurs before him, Shuman developed plans to ship the machine to North Africa. In order to buy property and move the machine there, he formed the Sun Power Company and funds were secured from a group of English investors. Before construction, the investors insisted that British physicist C. V. Boys was to review the machine and suggest possible improvements. Boys recommended a radical change; instead of flat mirrors reflecting the Sun onto a flat-plate configuration, he recommended a parabolic trough focusing sunlight onto a glass-encased tube.

With the suggestion by his own technical consultant, A.S.E. Ackermann, that to be effective the troughs would need to track the Sun, Shuman's desire to build a simple, easily constructed system was essentially gone. In 1912, the redesigned machine was built just outside of Meadi, Egypt; a farming community 15 miles (24 kilometres) from Cairo.

There were five collecting troughs, each measuring 204 feet (62 metres) in length and fitted with mechanical trackers that kept them facing the Sun (photo right). The Cairo plant outperformed the Tacony machine by a large margin, generating more than 55 horsepower and pumping over 6,000 gallons (27,276

Frank Shuman's solar powered pumping station in Meadi, Egypt, 1912.

litres) of irrigation water per minute. Shuman's fears over the use of complex mechanical systems proved to be unfounded and the exceptional performance of the Meadi plant made up for the higher installation costs. At a cost of $150.00 per horsepower, the machine could compete with fossil fuel systems and Shuman's company made plans to build more than 20,000 square miles (51,800 square kilometres) of reflectors in the Sahara desert.

This part of Egypt was a pre-World War One, German territory. Upon presenting his plans to the Reichstag, Shuman won a commitment equal to $200,000.00 towards his venture but just two months after the African machines' final trials, Archduke Ferdinand was assassinated in the Balkans, igniting the First World War. The fighting quickly spread to the upper regions of Africa and Shuman's solar irrigation plant was destroyed. Frank Shuman died before the armistice ending the war was signed and the plant was never re-built.

A History of Solar Powered Invention

Since the time of Frank Shuman, the remainder of the 20th century's solar design progress went without any truly new ideas in respect to those that have been detailed. What was to be learned about designing machines that could collect and utilise the heat in sunlight had apparently been learned and solar technology became commonly known, although still not commonly utilised. Though many more companies have formed to provide items ranging from household solar hot water heaters to actual community-sized power plants, only superior design and new materials technology now gives any application an advantage over its predecessors.

The Inventors and researchers highlighted in this chapter are just the tip of the iceberg in the history of solar thermal achievement. At the back of the book, in *Appendix One*, I have assembled a more complete timeline that, while a little more clinical in its approach than the historical narrative just completed, is a far more thorough record. The timeline was constructed from a seemingly endless range of sources an is, undoubtedly, still not the be-all/end-all authority on the record of solar thermal achievement. Just the same, it required a fair amount of time and effort to compile so hopefully it will delight readers that have a keen eye for such attention to details.

In wrapping up this section of the basic principles and history of solar thermal design, I would like to provide the following commentary; my own observations in regard to the use of the warmth found in sunlight.

I find it odd that the use of solar heat is, for some reason, still seen as a fringe technology by most people. In the mainstream, companies and individuals everywhere continue to burn things like wood, coal, petroleum products, animal fats and extracted plant oils to make heat for domestic and industrial purposes, while solar heated power continues to stay on the 'back burner', if you will. Why have the early discoveries and later adaptations and improvements not led to commonplace applications in our everyday lives? Why are people still fascinated when they realize that solar devices producing hot water produce water that *really is hot*. Solar powered devices work, they are easy to construct from relatively low-tech materials, many have low maintenance requirements and some of them have been with us for over two thousand years. It's enough to make poor Archimedes beat his head on the table.

We know there is a cleaner, truly cheaper and basically inexhaustible source of heat available to us in the Sun. We know that it can supply a wide range of our home and industrial heating needs, yet we choose not to take advantage of it. Perhaps it just seems to be far too simple for our prestigious position as the 'supreme beings' on this planet to rely on the Sun for heating. For those that think this is the case, perhaps the next chapters, which cover a range of sciences that need to be considered when studying solar heat, will shed some light on just how compex this simple source of heat truly is. Maybe with an understanding of the science involved, those taking the view that using the Sun for heat is primative or too basic wont feel quite so supreme as to turn their backs on what the Sun has to offer us.

The Science Behind the Sunlight

Chapter Three

It All Starts at the Sun

The Sun, nuclear science, Albert Einstein and the theories of relativity; what do they all have in common? Well, actually far more than I have room to write about in this little book! But we'll stick with the obvious answer given the topic that we're working with: solar thermal design. For without the Sun there would be no heat on our space bound chunk of matter that we call Earth and therefore we would have no need for solar heat because there would be no 'us' to have to keep warm and cozy or cook for. Thanks to Albert Einstein, his theories of relativity and wave-particle duality we are able to understand what is going on with the building blocks of the Sun and cosmos and therefore can make sense of the chain of events that produce the warming sensation of sunlight on our skin.

Knowing that solar thermal devices involved collecting **solar radiation**, in my early, passive solar learning phase I reasoned that I should start the search for understanding at the source – the Sun. Hmmm, it's a little too hot for me there … so I started at my Webster's dictionary and the word 'solar'.

Solar (sóulər) *adj.*– of or relating to the Sun.

That seems pretty straightforward but just to make things more interesting, let's ask the question "what is the Sun?"

The ancient Greeks called it Helios and later the Romans called it Sol; we all know it's that bright thing that we see up in the sky every day, but just what is it? Whatever name people use to refer to it, the Sun is a fascinating ball of gasses known to astrophysicists as a main sequence, G2 star. In the early 1900's, astronomical researchers Ejnar Hertzsprung from Holland and Henry Norris Russell from the United States developed a system for classifying stars in which a letter represents a range of surface temperatures and a number represents the luminosity (brightness) of a star as compared with other stars. The G group, which our Sun belongs to, refers to stars with a surface temperature between 4,000 and 7,000K (K represents the kelvin temperature scale, please see next page for details). The Sun's luminosity is rated at 2 on a scale going up to 1,000,000.

The Sun glows with a yellow-colored light and is one of more than 100 billion stars in our galaxy, the Milky Way. Although it is about 93,000,000 miles (149,600,000 kilometres) away from us, the light it creates takes only about eight minutes to reach our happy little planet and warm us up. Why does sunlight warm us? We'll be exploring that phenomenon in a couple of different places in the book, with the first in just a few more pages.

Here are a few vital statistics about the nearest star, our Sun:

- **Age:**   4.5 billion years old

- **Diameter:**   864,000 miles (1,390,000 kilometres)

- **Mass:**   1.989e30 kgs. This figure is using scientific exponential notation to save on a ridiculous amount of zeros. "e30" means to multiply 1.989 by 10 to the 30th power or: 1,989,000,000,000,000,000,000,000,000,000 kgs.

- **Temperature:**   5,800 K (surface) 15,600,000 K (core)

- **Composition:**   about 70% hydrogen and 28% helium (by mass). Other compounds, basically metals created by the Sun's internal processes, amount to less than 2% of its total composition.

Two temperature scales are commonly used in our everyday lives and industry, Fahrenheit and Celsius. There is however a third temperature scale that is more common to science, known as the kelvin scale. What is the difference between these scales? Let's find out now.

Fahrenheit (°F) is a temperature scale in which the temperature difference between the melting and boiling points of water is divided into 180 equal intervals, called degrees. The freezing point is 32° and the boiling point is 212°. This scale was established by the German-Dutch physicist Gabriel Daniel Fahrenheit in 1724 and although the scale was formerly used widely in English-speaking countries, many have changed to the more convenient Celsius scale. An unfortunate and notable exception is the United States, where the Fahrenheit scale is still in common use along with Imperial units of measurement. Temperatures on the Fahrenheit scale can be converted to equivalent temperatures on the Celsius scale by first subtracting 32° from the Fahrenheit temperature, multiplying the result by five and then dividing the result by nine. The written formula is: $(F-32) 5/9 = C$.

Celsius (°C) is a temperature scale according to which the difference between the freezing and boiling points of water is divided into 100 degrees. The freezing point is 0° and the boiling point is 100°. The scale is named for the Swedish astronomer Anders Celsius, who established it in 1742. The scale is also referred to as the centigrade scale because it is divided into 100 degrees (centi from the Latin centum, meaning 100). In 1948 the 9th General Conference on Weights and Measures changed the official name from centigrade degree to Celsius degree, so centigrade is officially an incorrect term. Temperatures on the Celsius scale can be converted to equivalent Fahrenheit temperatures by multiplying the Celsius temperature by nine, dividing the result by five and then adding 32° to that number. The written formula is: $9C/5 + 32 = F$.

The **kelvin (K)** temperature scale is named after the British mathematician and physicist William Thomson Kelvin who proposed it in 1848. Kelvin has an absolute zero temperature below which colder temperatures do not exist; this is equivalent to minus 273.15° on the Celsius scale and minus 523.4° on the Fahrenheit scale. The Kelvin degree is actually the same size as the Celsius degree, so the reference temperatures for the freezing point of water (0°C), and the boiling point of water (100°C), correspond to 273.15K and 373.15K. Temperatures on this scale are called kelvins, not degrees kelvin and the word kelvin is not capitalized. In 1967, the 13th General Conference on Weights and Measures decided that the symbol 'K' (which is capitalized) stands alone with no degree symbol when writing temperatures in this scale.

It All Starts At the Sun

Aside from the obvious fact that the Sun gets older each day, the rest of the Sun's statistics (diameter, mass, temperature, composition) are also changing slowly over time. The Sun is getting brighter, hotter, larger and its composition is altering as a result of the chemical and atomic processes happening internally. From its visibly bright skin to the depths of its hidden core, since forming some 4.5 billion years ago, scientists estimate that the Sun has used up about half of the hydrogenthat it originally contained. Fortunately for us, the Sun has a lot of hydrogen left; it will continue to keep ticking along by turning hydrogen into helium through a process known as **nuclear fusion** for about another 5 billion years or so.

While this time passes by, the hydrogen that the Sun is using for fuel will begin to run out. When this happens the helium, present in ever growing quantities from hydrogen fusion, will begin to fuse into carbon. The light from this 'new' version of the Sun will become red in color and nearly double in intensity from what we currently experience. While the Sun slowly goes through this process of turning into what is known as a red giant, it will enter the 'K' classification of stars (as opposed to the 'G' class already mentioned) and its size will increase to a point around where that of where our good 'ol Earth currently resides. Sorry but its true, the expansion of the Sun will eventually result in the total destruction of the inner planets of Mercury and Venus, quite likely Earth and possibly the next farthest planet past Earth, Mars.

The expansion of the Sun will however someday run out of momentum and at that point it will begin to collapse. The Sun will have exhausted its supply of commonly fusible elements and if it follows a typical star life cycle, in order to stay alive it will eject its outer layers and become a white dwarf, reducing to perhaps $1/50^{th}$ of its current size (about the same diameter as that of our planetary neighbor, Jupiter). White dwarf stars are remarkably dense bodies; a portion of a white dwarf the size of this book may weight thousands of tons. With that density comes a powerful gravitational field, so any planets left over from our current solar system will very likely be pulled in and consumed by the little star. If they manage to escape that fate, their orbits will then probably become larger than they currently are and they may even slip away into open space.

If the Sun does not follow this standard life cycle other fates may await it. A black hole could wander into the general area of our solar system and then Sun, along with everything else in the solar system, could be consumed by it. Perhaps through a series of cataclysmic events, the Sun could explode and become a supernova … who knows? Hopefully those two scenarios won't present themselves while we're around.

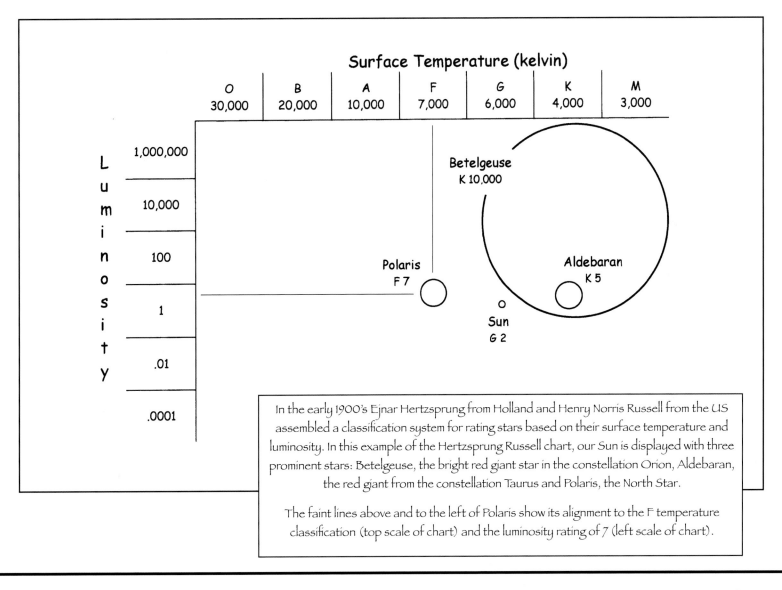

Surface Temperature (kelvin)

O	B	A	F	G	K	M
30,000	20,000	10,000	7,000	6,000	4,000	3,000

L
u
m
i
n
o
s
i
t
y

1,000,000

10,000

100

1

.01

.0001

Betelgeuse
K 10,000

Polaris
F 7

Aldebaran
K 5

Sun
G 2

In the early 1900's Ejnar Hertzsprung from Holland and Henry Norris Russell from the US assembled a classification system for rating stars based on their surface temperature and luminosity. In this example of the Hertzsprung Russell chart, our Sun is displayed with three prominent stars: Betelgeuse, the bright red giant star in the constellation Orion, Aldebaran, the red giant from the constellation Taurus and Polaris, the North Star.

The faint lines above and to the left of Polaris show its alignment to the F temperature classification (top scale of chart) and the luminosity rating of 7 (left scale of chart).

It All Starts At the Sun

Except at their cores where extreme compression forces exist as a result of a star's massive gravity, they are far too hot to maintain a structure made up from anything but gases. As a result of being a rotating ball of gas, the outer layers exhibit a fascinating characteristic known as **differential rotation**. At the Sun's equator, the surface rotates once every 25.4 days but near its poles the rotation can takes as long as 36 days; of course between these two points the surface rotation is literally any number of days in between. The differential rotation effect extends well down into the interior of the Sun but the core (approximately the inner 25% of its radius) rotates as a solid body. Interestingly enough, a similar differential rotation effect has been noted in our solar system's gas planets, Jupiter, Saturn, Neptune and Uranus.

While figures vary slightly when consulting a range of sources, the Sun apparently contains more than 99.8% of the total mass of our Solar System. Although it is relatively small in size when compared with other stars, when measured by its mass against the general group of stars in our Milky Way Galaxy, it rates in the top 10%. Research on stars other than our own is far from complete, but scientists suggest that although generally larger, an 'average' star in our galaxy contains less than half the mass of the Sun.

Rising now from the interior and turning our attention to a star's outer layers, the surface of a star is called the **photosphere**, which in the case of the Sun is a relatively thin layer about 65 miles (100 kilometres) thick. As mentioned back in the statistics, the photosphere (surface) of our nearest star is operating at a temperature of about 5,800 K. There are anomalies to this generalisation though, called **sunspots**. They are called sunspots because when observing the Sun through a filtered telescope they look dark when compared with the regions surrounding them. A single spot can be found to be covering as much as 30,000 square miles (50,000 square kilometres) of the solar surface and it turns out that these big, dark spots are also cool spots. They are found to be around 3,800 K – 'cool' in this instance being used in the ultimate relative sense. Scientists do not fully understand the processes that cause sunspots very well, but interactions of the photosphere's surface matter with the Sun's magnetic field are high on the list of reasons for their appearance.

Moving this investigation of star anatomy off of the surface, there is a small region known as the **chromosphere,** a reddish colored area about 3,100 miles (5,000 kilometres) thick that lies above the photosphere and then there is a rarefied (low density) region above the chromosphere, called the **corona**.

The normal light of the Sun blinds us to the corona but if you're ever lucky enough to be in a region of a solar eclipse, it is visible during the period that the Moon is fully obscuring the Sun from view. Looking like an eerie halo of flames, temperatures in the corona soar to over 1,000,000 K and it extends several million kilometres into space beyond the solar surface.

Photo Credit : SOHO (ESA & NASA)

Two instruments aboard the SOHO spacecraft sent this composite image of the Sun. UVCS imaged the outer coronal region and EIT the inner portion that shows the solar disc. Dark area s called coronal holes, where the highest speed solar wind originates, can be seen at the poles and across the disk of the Sun.

It All Starts At the Sun

There's something odd about that; how can the temperature of the corona be greater than the temperature of its source, the photosphere? The intense heat found in the Sun's corona is a scientific mystery because it cannot be ordinary heat. Ordinary heat can never create a region with a higher temperature than its source. While theories ranging from raining meteorites to magnetic field induced plasma have been explored, the corona's high temperature is still an unexplained phenomenon.

We can see the light from the Sun and we can feel heat coming from it, but the Sun also emits things that we can't see or feel unless conditions are just right. A low density stream of charged particles that comes from the Sun is known as the **solar wind**. Solar wind is made up mostly of invisible particles breezing throughout the solar system at about 280 miles (450 kilometres) per second – some breeze! There are also high density particles that burst forth from the Sun creating **solar flares**. If we're back on the filtered telescopes and looking at sunspots at just the right time, we might be so lucky as to also see a solar flare, also called a solar prominence, blasting from the surface like a massive lick of flame reaching out into space.

The low-energy particles that are emitted as solar wind, and the much higher energy particles ejected by solar flares, can have dramatic effects back here on Earth. While we can never see the charged particles racing past us or crashing into the planet, astronauts working above the Earth's protective atmosphere report seeing flashes of light when they close their eyes. Some researchers believe that the flashes are a result of these charged particles impacting upon the retinas in the astronaut's eyes. Back here on the surface, under the protective blanket of the atmosphere, the particles can manifest in such nuisances as power line surges and radio interference or such glories as the hauntingly beautiful aurora borealis of the northern polar regions and the aurora australis of the southern polar regions.

EIT 304Å1 (a scientific instrument aboard the SOHO spacecraft) sent this image of a huge, handle-shaped solar prominence on Sept. 14, 1999.

Prominences are huge clouds of relatively cool, dense plasma suspended in the Sun's hot, thin corona. At times, they can erupt, escaping the Sun's atmosphere.

In this image, every feature traces magnetic field structures. The hottest areas appear almost white, while the darker red areas indicate cooler temperatures.

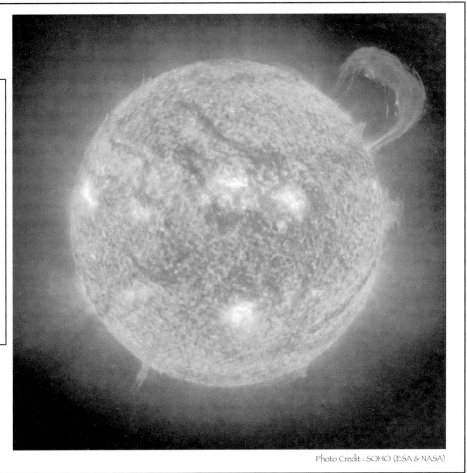

Photo Credit : SOHO (ESA & NASA)

It All Starts At the Sun

Whew! That takes care of the basic knowledge that I have to offer about the Sun but before proceeding on to further discussion about solar heated devices, we still need to know what the 'radiation' part of the term 'solar radiation' is. I will start this once again back at my Webster's dictionary; this time with a look-up for the word **radiation**.

Ready? Here goes …

> **Radiation** (rẹidi:éiʃən) *n.*   the act or process of radiating; energy radiated in the form of waves or particles; loss of energy by the emission of waves or particles.

In my own learning quest, I can still remember the first time that I read that definition. No problem back there with the definition of the word 'solar' but I must admit, while I immediately considered how very impressive I would sound while quoting the Webster's definition of radiation to friends that wouldn't be willing (or interested enough) to ask for further details, I didn't think that it gave me a very firm grasp of what radiation was or how it worked. Honestly, I just sat there bewildered, thinking, "what the heck … ?" So I chose to find a better answer by spending an afternoon in the physics section of the local library.

Rows and rows of books covering the subject of physics confronted me at the library. Stepping over to the look-up computer, I thought to myself, "where do I start?"

At least I like the musty smell of old study books and physical science is a sideline interest of mine; I would most likely enjoy whatever I would learn, even if by some chance it didn't end up including a firmer understanding of radiation. I inputted "radiation" for the keyword search criteria on the library's computer and hit the enter key.

Wow, what a great afternoon I had. With the help of all of the lovely smelling physics books that I read through, I learned that solar radiation is a by-product of the nuclear fusion processes happening at the Sun's core. Astrophysicists tell us that at the center of the Sun there is so much pressure (some 250 billion times more pressure than what we feel standing on the surface of our home planet) and so much heat (15.6 million kelvin) that **atoms** are crushed together until, in some instances, they form new compounds, that is, they are *fused* into new compounds. It is from this process of nuclear fusion that solar radiation is produced.

In the midst of all this heat and pressure, when nuclear fusion happens an interesting phenomenon occurs; it appears that the newly created atom is lighter in mass than that of the atomic building blocks that went into creating it. What does that mean? Believe it or not, in nuclear fusion:

1 + 1 does not equal 2.

The most well-known fusion reaction happening at the core of the Sun is that of two hydrogen atoms fusing together and becoming a helium atom. Due to the influence of a force in the nucleus of an atom called the **strong force**, the weight of a helium atom is actually less than the sum of the weight of its components. Now, *that* I found interesting.

What does all this have to do with radiation?

What is an Atom?

And what does all this have to do with solar thermal designing?

Please read on … all will be revealed in the next chapter, *Nuclear Science Basics.*

It All Starts At the Sun

Explore the Universe

If the urge ever strikes to have a look at the Sun, say during an eclipse or perhaps just to see if you can see any sunspots, please don't look directly at it with your naked eye or, even worse, through an unfiltered magnifying device like binoculars or a telescope. I know that sounds like a silly warning, after all, our parents have probably told most of us this since before we can even remember but the Sun can cause permanent eye damage in a matter of seconds that cannot be felt as it is happening. I don't want to be responsible for getting someone interested in the Sun and then having them burn their retinas out, never to see anything again.

One way to observe the Sun that works well and costs little is to use a pair of binoculars as a solar projector. Pointed at the sun with a piece of white paper placed behind the eyepiece, the Sun's image is easily projected onto the paper by the intense, magnified light coming through the binoculars. A common pair of 7 x 35 binoculars are good enough to clearly detect sunspots using this method.

For those keen to step up their financial investment, a wide range of telescopes equipped with Sun filters are readily available and for those that wish to spend as little as possible to satisfy their Sun-viewing curiosity, a piece of cardboard with a pinhole can be put to work by projecting the sunlight onto the ground after it passes through the pinhole. The only issue with this method is that there is no magnification; you won't see much other than a bright spot of light on the ground … but that bright spot of light is the Sun.

So let's say you can't afford binoculars or a telescope and the spot of light on the ground is just not very exciting; you can't stand the curiosity and you have to look directly at the Sun. Got any exposed 35mm film around the house? For exposed black-and-white film, four or five layers of the completely black part (where there are no photos – just black) can be used as a filter for direct viewing. I like to use layers of these dark parts of exposed color film, because it gives the solar disk an orange color that my mind equates with the Sun. To go really up-market for direct Sun viewing use a welder's shield, but now we're back to spending money. Whatever method used in direct viewing, if the Sun isn't comfortably dim immediately look away.

Chapter Four

Nuclear Science Basics

One sure way to kill an average dinner party conversation is to choose particle physics as a subject. Not to be shy, I sometimes find that the topic will get a bite from a few players in the crowd, but usually the blank stares can feel quite intimidating, just before conversation settles back to detailed analysis of the weather and other very critical points from the daily news. Those that do fancy a dabble or more in the subject will often get caught up in analysing how something's 'parts' contribute to its 'whole'. It is after all, the parts of something that in the world of physics are the particles.

Most people know how to use a hammer and although hammers usually don't have anything to do with particle physics, lets suppose for a moment that we could pull out the most special hammer ever devised, one that would allow us to put a piece of 'something' onto a workbench and smash it into its most basic parts. While banging away, we would see the initial piece of 'something' turn to chunks, to chips, then sand, then dust and eventually, getting past the visible pieces yet still having a piece of whatever our 'something' was, we would get down to a point where we would have a **molecule** of it. Molecules can be described as "the smallest amount of a chemical element or compound that can exist while still being that element or compound".

After having discovered molecules, we might wonder if there was anything smaller than a molecule, What made up our "smallest amount of a chemical element or compound that can exist while still being that element or compound"? To find out, we simply need to do a little more work with our special hammer.

If we keep hitting the molecules with the special hammer, going farther in our quest to completely disassemble our 'something', an additional blow or two will reveal that molecules are made from **atoms**, which are defined as 'the smallest portions of matter that display the characteristic properties of a particular element'. That definition doesn't sound very different from the definition of a molecule, but remember that we are dealing with some very tiny considerations here. The key learning point is that atoms make molecules.

Can we smash atoms into something smaller still, with our special hammer? Indeed we can.

With another well-directed 'whap!' from our special hammer, we can break atoms down into even smaller bits called the **fundamental particles**. The fundamental particles are the **neutron**, the **proton** and the **electron** and they are what atoms are made from.

Are we to the end yet?

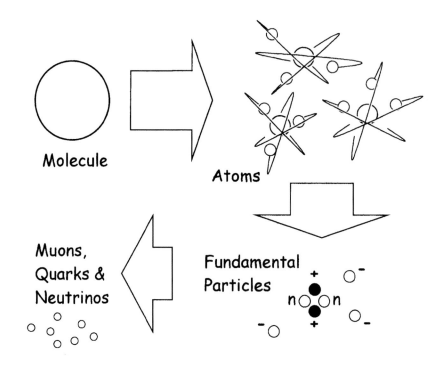

Molecule

Atoms

Fundamental Particles

Muons, Quarks & Neutrinos

For our purposes, yes we have reached the end – the actions of fundamental particles are all that we need to consider when talking about solar radiation but for the needs of the universe, no; we haven't. The incredibly special hammer can break these fundamental particles into even smaller pieces; **quarks, neutrinos and muons.**

Keep in mind that we humans haven't quite managed to split the fundamental particles down into quarks, neutrinos and muons yet. In spite of the researcher's real hammers (devices called particle accelerators) not being as good as our 'special hammers', physicists have other ways of knowing that they're there. Truth be told, given enough time, along with big enough minds and/or computers and science may discover that there are even more small particles to be found beyond the quarks, neutrinos and muons.

If we stepped back one hammer blow and stopped with neutrons, protons and electrons we would have the basic building blocks that are required for making up atoms. That means we also have the atomic pieces needed to make the fusion reactions found within the Sun.

Fusion and Radiated Heat Transfer

Let's think about magnets; big, red U-shaped ones with silver ends that were used in my high school science class come to my mind. One end of the U has a positive magnetic polarity and the other end of the U has a negative magnetic polarity. The way that magnetic polarity works appears somewhat similar to electrical charge, which is also expressed as positive and negative. What is an electrical charge? No one knows, but we do know that protons have a positive charge and electrons have a negative charge; neutrons are ... you guessed it, neutral. Like magnets in a high school science class with opposite charges, a proton and an electron will attract each other. Two protons (or two electrons) that have the same charge will, just like magnets of the same charge, push each other away, although different natural forces are involved.

If we were to decide to build an atom, we know from our 'special hammer' experiment that we will need some fundamental particles, again: protons, electrons and neutrons. The blueprint for an atom is pretty straightforward consisting of a core, called a nucleus, which is made up of protons and neutrons and a cloud of electrons that surround the core. This sounds pretty easy, but a sustainable, controlled fusion of elements is something else that humans have not yet managed to achieve. That's probably just as well, at least for the time being, as the result of fusing elementary particles together is a massive burst of energy that, although it would be completely able to provide for the power needs of all Earth's inhabitants, it also just might cook us like a microwave oven that had been left open (for an example of human-induced nuclear fusion, think of hydrogen bombs). So if we can't manage fusion here on Earth, how does it happen in the Sun?

Nuclear Science Basics

Remember all that heat and pressure at the core of the Sun that was discussed in the last chapter? The intense heat and extreme pressure are vital elements in the fusion process but the final link, the piece that brings it all together, is yet again an unexplained force of nature. Imagine for a moment a force capable of drawing together the protons and <u>neutrons</u> that is 100 times stronger than the electromagnetic attraction/repulsion that the protons and <u>electrons</u> already experience …

Now we're ready to talk about nuclear fusion and radiation.

Nuclear Fusion

When atomic particles join to form a new element under the conditions present in the Sun, the oppositely charged protons and electrons lump together and pack in very tightly as a result of the attraction they feel from electromagnetic forces. Of course, that ever-present crushing force of 250 billion Earth atmospheres in the center of the Sun helps them to pack in as well and in these conditions they will tighten up until an element of electrostatic repulsion between particles with like charges also begins to take place, the two forces creating a state of magnetic equilibrium. For any given element, the necessary numbers of neutrons (and this number can vary, even for the same element) also nestle in with the protons and electrons.

This would be the final resting position of the particles if left on their own, but this just happens to be when the strong force comes 'alive' and takes over. Literally for the last thousandth of a trillionth of a metre the strong force pulls the protons and neutrons together even tighter, finally forming the powerful bond of an atomic nucleus, while the electrons begin to orbit around them. So what happens to the mass that was present in the form of the **electrical charges** represented by the attraction and repulsion forces? It is forced out, or given off (as the physicists say) in the form of … **electromagnetic radiation.**

$E = mc^2$ *rules!*

The strong force is so powerful that the nucleus of a helium atom has about 99% of the mass present when its individual elements were part of hydrogen atoms. When dealing with nuclear fusion, 1 + 1 = 1.98.

This diagram shows the basic process of nuclear fusion. Item one depicts the attraction between protons and electrons, and the pressure and heat present at the core of the Sun.

In items two and three, the particles represented are the neutrons and protons of the atomic nucleus as they go through the process of nuclear fusion, forming a new compound and releasing excess energy that becomes electromagnetic radiation.

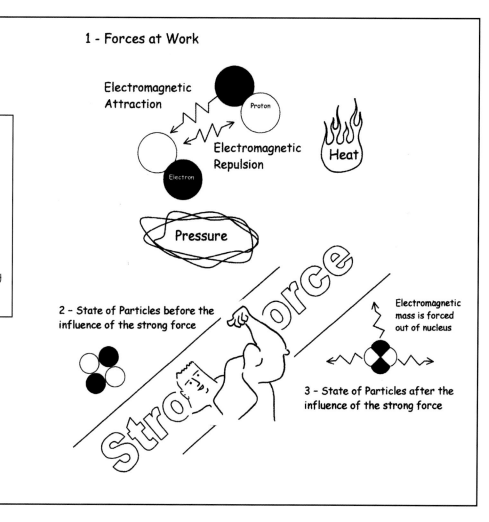

1 - Forces at Work

Electromagnetic Attraction

Proton

Electromagnetic Repulsion

Heat

Electron

Pressure

2 – State of Particles before the influence of the strong force

Strong Force

Electromagnetic mass is forced out of nucleus

3 – State of Particles after the influence of the strong force

Nuclear Science Basics

Each and every second of the day about 700,000,000 tons of hydrogen is converted to about 694,000,000 tons of helium at the core of the Sun. As a by-product of the fusion process that turns the 700,000,000 tons of hydrogen into helium, about 6,000,000 tons of energy, in the form of electromagnetic waves called **gamma rays**, is given off. Gamma rays are emitted from the helium atom's nucleus at the point when the strong force exerts its influence and pulls the protons and neutrons together.

The force of the Sun's gravity is incredibly strong and as the released energy travels outward, toward the surface of the Sun, it is continuously pulled back, absorbed and re-emitted at lower and lower levels as it gets closer to the surface. By the time it reaches the surface, the energy has been reduced primarily to the visible light that we see. Sunlight is energy in the form of light, which has managed to escape the extreme conditions present at the Sun.

Through processes that we will discuss in greater detail later in the *Nature of Materials* section, that light is converted into what we humans feel here on Earth as warmth, like what we feel on our face when we stand outside on a sunny day. So when we feel the warm sunlight on our face in spite of chilly, springtime air, this is what is happening – first at the Sun and then here on Earth:

- Energy is created through the fusion of hydrogen and other elements at the Sun's core;

- The energy is emitted through a stream of magnetic and electrical particles known as electromagnetic waves;

- Solar radiation (the electromagnetic waves) strikes the surface of our skin and is then converted to heat;

- The heat activates the molecules that are contained within us and warmth spreads through our bodies.

With an understanding now of what creates solar heat at its very source, we are well equipped to begin exploring the properties of heat that are so vital to designing effective collectors of solar radiation for the purpose of capturing and storing the warmth in sunlight. The information that we seek about how heat acts is found in the science of heat, a field of study known as **thermodynamics**.

Chapter Five

Principles of Heat in the Universe

Thermodynamics is a branch of physics that studies how groups of bodies that have a temperature interact with one another. This is another great topic not to bring up at dinner parties, but funnily enough it impacts on all of us nearly every day. Have you ever felt cold? That feeling is basically the air around you trying to warm itself with your body heat. Have you ever felt blistering hot water come from a garden hose that had been lying on the lawn under a sunny sky? That is the water in the hose retaining heat that the hose is receiving from solar radiation. In both situations there was at least one body giving heat and one body taking heat, thus the basic setup for thermodynamic study was present.

Thermodynamic science calls such groupings of temperature bearing bodies a **system**. Thermodynamic science is based on the concept that in an isolated system anywhere in the universe, there is a measurable quantity of energy. This energy within a system is called the **internal energy** of the system and systems must involve the flow of energy from one body to another. Systems include objects from the very large (stars) to the very small (molecules and atoms) and thermodynamics recognizes that energy of all kinds can be transferred directly as heat, therefore this branch of science excludes chemical and nuclear exchanges, instead focusing strictly on the heat aspect of system interactions.

Heat energy is present in a variety of states and the energy of bodies in motion is called **kinetic energy**. An object that has motion in any direction, be it vertical, horizontal or somewhere in between is displaying kinetic energy. There are three main forms of kinetic energy:

Vibrational - energy due to vibrating motions;

Rotational - energy due to rotating motions;

Translational - energy due to a body moving from one location to another.

Like kinetic energy, **potential energy** also appears in more than one form but bodies displaying potential energy aren't actually doing anything … yet. The **gravitational** potential energy of an object here on Earth is the energy stored in a body as a result of its resting height above the ground. This position gives the body a relationship with the gravitational attraction of the Earth in that the higher that an object is, the greater the gravitational potential energy. There is also a direct relationship between gravitational potential energy of a body and its mass; the more massive the body is, the greater its gravitational potential energy is here on Earth.

The other form is of potential energy is **elastic** potential energy. Elastic potential energy is the energy stored in materials that are capable of being stretched or compressed. Elastic potential energy can be stored in rubber bands, the tendons within our bodies, and anything else that might have stretchy or squishy properties. The amount of elastic potential energy stored in such a body is directly related to the amount of stretch or squeeze placed upon it – the more the body is stretched or squeezed, the more stored elastic potential energy it has.

When the properties of a thermodynamic system are considered, the study usually includes the total of both the kinetic energy and potential energy of the system. This total is the internal system energy and it is represented in thermodynamic calculations by the letter "U".

Systems under consideration in thermodynamic studies are usually isolated, therefore the value of U (internal system energy) can only be changed if the system ceases to be isolated; it must either receive or give energy from or to something (or someone). That energy exchange can be the transfer of mass to or from the system, the transfer of heat to or from the system, or work being done by the system. The symbols used in scientific equations for these values are as follows:

ΔU = change in internal system energy;

Q = heat added;

W = work done by the system.

Now we're starting to talk about thermodynamics!

The study of these energy transfers has resulted in a series of laws that govern the conversion of energy from one form to another, the direction in which heat will flow and the availability of energy to do work. In its most basic form, the **First Law of Thermodynamics** is:

energy is conserved

The law is written out as: $\Delta U = Q - W$ and demonstrates that any energy transfer between systems will change the internal energy of all systems involved. Also, at the end of a transfer the total energy of the systems involved remains constant; the change in the internal energy of two systems is equal to the heat added to the system, minus the work done by the system; what one body gains, another loses in equal proportion.

The law as it applies to two systems interacting would therefore be written out as:

$$\Delta U =$$

$$(Q - W) \quad + \quad (W - Q)$$

(System 1 – heat is added and work is done)   (System 2 – work is done and heat is lost)

To visualize this law, let's use a rock up on the top of a hill, as shown in the diagram on the opposite page. It is displaying some level of potential energy due to the nature of its mass and its elevated position. It isn't doing anything while sitting there, but it does have the ability to do something if it is set in motion to go rolling down the hill; the potential of what it may do is held within.

Think about that rock on the hill. Its potential energy can be seen in the equation as the value U. Now, to get it rolling something or someone has to give it a push and get it started; the energy that is given to the rock by that something or someone is represented by the value Q in the equation. An amount of energy that is equal to the energy given to the rock is lost from the giver and in this part of the equation we could find the change in both internal systems; ΔU.

Finally receiving a push from a mischievous youngster, the rock tumbles down the hill, careens through a fence and comes to a rest on top of an unsuspecting rabbit that was sitting inconveniently in its path; we can say that the rock did some work! What is work in the equation? ... W.

In order to come to a stop, the rock transferred kinetic energy into the ground as it rolled, the fence as it crashed through and the rabbit as it finally comes to a rest. These bodies, along with the youngster that set the rock in motion would make up the basic set of bodies within the system being studied. In this example, the object's potential energy was converted to kinetic energy when it was set in motion, while the eventual interaction with other bodies conserved the total energy of the developing system by converting it back into potential energy. What was gained by the fence, the ground and possibly the rabbit (well, the rabbit's mass anyway) was first lost by the person pushing the rock and then the rock as it rolled along.

In this manner, internal energy within a system can be converted between potential energy and kinetic energy. It can subsequently be stored within the system, however we must remember that, as the law shows, energy depends upon bodies in a system both giving and getting for the energy to be conserved. As can be imagined, the first law of thermodynamics allows for many possible states of bodies within a system to exist, although we only see a few naturally occurring here in our solar system.

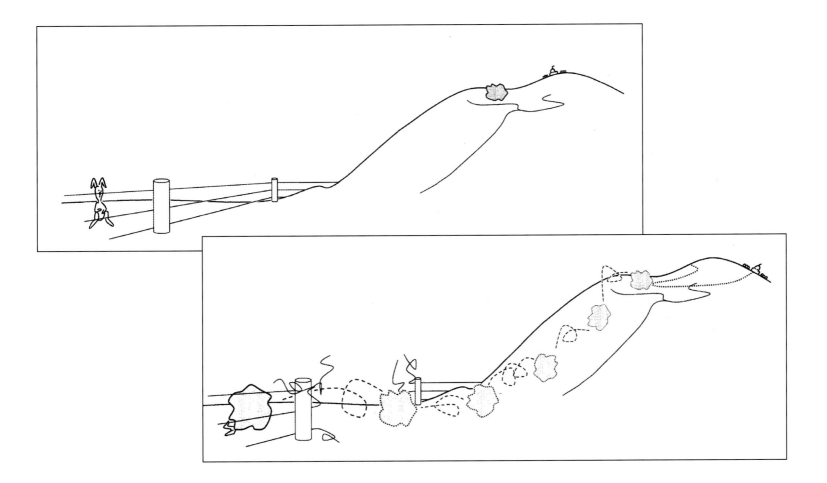

Now, to relate this thermodynamic study to solar designing concerns: looking to the end of the last chapter I made a point about sunlight striking our skin and converting to heat – this point is important. We must remember that before the surface contact is made, solar radiation is really only a stream of electromagnetic waves flying through space. The presence of matter is not required for energy to transfer by radiation and therefore it can travel through the cold vacuum of space, spanning the distance from the Sun to our planet in this form.

Since the time of the renowned scientist Sir Isaac Newton (1642–1727) a major dilemma in the study of physics was how to explain the nature of heat transfer by radiation. The topic puzzled scientists for centuries and continued to present problems right up to the late 1800's, when many individuals were trying to define the laws of the then relatively new science of thermodynamics.

James Kirk Maxwell (1831–1879) developed a theory in which kinetic energy was reasoned to be at the core of radiant heat transfer. Maxwell proposed that this form of energy travels as a vibratory electric and magnetic disturbance through space in a direction perpendicular to those disturbances. This theory fit in nicely with the concept of vibratory kinetic energy and the basic laws of thermodynamics where heat is released by a body (the Sun) and received by another body (say, our planet). In a simple description of his work, Maxwell envisioned that molecules making up a 'perfect gas' would move about and collide with each other, like balls bouncing off the surface of the container holding the gas. As we know, the energy associated with motion is kinetic energy and this kinetic approach to the behavior of perfect gases led to an interpretation of the concept of temperature transfer on a microscopic scale – including radiation. The concept explained radiation as a form of kinetic heat transfer. Eventually this work spun off the scientific field of molecular dynamics.

However, to cut what could be a very long story short, James Maxwell wasn't exactly correct and it took a later scientist from Germany, Max Planck (1858–1947), to sort out the dilemma of radiated heat transfer by developing factors such as Planck's Law, the Planck Constant and the Planck Curve along with theories supporting the idea that the energy of the oscillating (vibrating) electrons of atoms present at the surface of the radiating object must be quantized. Yikes! While most of Planck's work is WAY beyond my level of interest, it leads to the theory of **blackbody radiation**, which is very important to consider when dealing with heat transfer by radiation (a brief discussion on blackbody radiation is coming up in *Section Three – The Nature of Materials*).

Max Planck, one of the most significant contributors to an understanding of electromagnetic radiation and the field of thermodynamics.

Lets face it ... he'd probably be asleep by now! Given the magnitude of his discoveries, I really doubt he would find much in this book that would keep him engaged.

Consider the following principle of energy in the universe as the **Second Law of Thermodynamics**:

Energy tends to disperse towards a more even distribution

This law is manifested by the tendency of heat to travel from high temperature areas to low temperature areas in an effort to disperse. The definition of the laws includes the term 'tends' because heat can occasionally move from a cold area to a hotter area and it can stay contained, although this seldom occurs without some form of outside influence; some form of applied work. Why are these circumstances unusual? Think for a moment of the fact that the presence of heat is actually an indication of the movement of atoms and heat dispersion is a representation of colliding atoms passing along their energy as they bump into one another.

The entire universe is made up of moving atoms; air, space, even solid objects … everything. We can usually visualize solid objects easily, so for this example I will use a block of metal receiving heat via radiation on just one side. That side of the object is hot, its atoms are moving quickly. The other side of the object, the one not receiving direct sunlight, is cold; its atoms are moving slowly.

In this scenario, where is there a greater opportunity present for the heat to disperse, within the volume of atoms on the side of the object that is hot (where the atoms are already moving fast) causing them to move even faster or by bumping into the slow moving volume of atoms on the cold side, causing them to begin to speed up?

Heat generally moves from high temperature areas to low temperature areas because there is simply more opportunity for fast moving atoms to transfer energy to slower moving atoms than by trying to give even *more* energy to other, fast moving atoms. So now you know, when someone in the house (other than ourselves, of course) leaves the freezer door open and the ice cubes all melt, it is most likely because heat comes in, not because the cold goes out.

Planck's work confirms that, although involving different processes, radiated energy transfer acts in a similar fashion to that of direct contact, energy transfer (that we're getting to soon) and that the only requirement for heat transfer via radiation is that the waves originate from a source that is hotter than the object receiving them. This of course is obviously the case between the Sun, operating at a temperature 5,800 K, and the Earth, which is operating at 300 K, or our bodies at 310.15 K.

Through dispersal, heat is constantly seeking a state of equilibrium, where all bodies in a system are the same temperature. Heat doesn't want to stay 'locked' in a warm object, it wants to find a cooler object to warm, either through radiation or other direct contact transfer methods. The state of equilibrium occurs when objects are both receiving and emitting heat at a constant rate. The objects will attain the same temperature, and in the absence of additional loss to other objects, they will then maintain a constant temperature. Thermal equilibrium is the subject of the **Zeroth Law of Thermodynamics**:

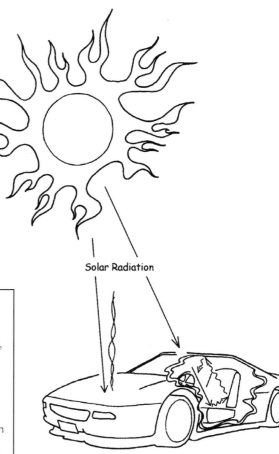

Solar Radiation

If two systems are in thermal equilibrium
with a third system, then they must be in thermal equilibrium
with each other

That seems pretty obvious really, since three systems in thermal equilibrium with each other will undoubtedly have the same temperature. Some reference books dispute the need for this to even be a law of thermodynamics, however it does present itself often enough that I've decided to include it.

Cars parked outdoors provide an excellent example of what happens when solar radiation strikes a surface, which then becomes hot.

Hot cars also demonstrate how radiation can easily penetrate clear glass, but after striking surfaces within the car it is trapped as heat; a classic example of the "Greenhouse Effect".

Direct Contact Heat Transfer

Once the surface of an object has been contacted by solar radiation, the object will begin to warm. Depending on the composition of the object's material, one of two methods of heat transfer will come into effect to warm the whole object rather than just its surface. If the object is a solid, **conduction** will allow the heat to move through it. If the object is a fluid or a gas the heat can travel through it by **convection**.

Conduction

To grasp the concept of conduction, a person has to think small but not quite as small as with radiation. Conduction actually happens at the molecular level of a solid object. Molecules are in a state of constant vibrational motion and heating will make them move around faster, cooling will make them move more slowly. It is this molecular motion in response to heating or cooling that is at the root of the conductive process. Heat moving through an object by conduction does so one molecule at a time. When a molecule becomes heated and begins moving faster it will bump into other molecules that are around it. When it does this it transfers some of the newly gained heat to its cooler friends as it bumps into them. These newly heated molecules will then begin to move around more than when they were when cool and transfer heat to other, colder molecules in the areas around them. This process will continue to happen, the heat gradually penetrating the solid object molecule by molecule, until eventually the object is brought to a uniformly warmed state.

Convection

Convection can occur only in fluids and gases because for convection to occur there must be free movement of molecules. Not just their normal, continuous vibrational movement, which is so small that it is almost on an atomic level. For convection the movement must be on a human scale, molecules able to move inches, feet or even miles.

When molecules in a fluid or gas are heated they expand significantly, becoming lighter in weight as a result of becoming less dense (maintaining the same molecular weight but occupying a larger space). Since molecules in a fluid or gas aren't solidly bound to other molecules around them, they will rise when they become lighter.

This brings us to another scientific principle, this one is in regard to the activity of matter while in the presence of gravity, such as when it is located on a planet:

Light elements rise, heavy elements descend.

Through the rising and falling of light and heavy molecules, heat moves through water, air and other liquids and gases. Convection currents of hot and cool molecules will flow throughout the liquid or gaseous body, as in the solid object, molecule by molecule until eventually the receiving and emitting of heat brings the entire mass to a uniformly warmed state.

The rising and falling of the liquid or gaseous molecules as they are heated, sets up the process of a thermosiphon. As the heated (therefore lighter) molecules rise, the space they had occupied is taken over by cooler, heavier molecules; as we know, they must fall. If this process is contained, say by having a liquid such as water trapped in a closed container, the water will begin to circulate on its own. Hot water goes up to the surface, cold water comes down to the bottom of the container. There are no requirements for electricity, motors or any other devices. With the correct combination of elements in place, it just happens.

Convection

When molecules are heated, they expand, and thus become lighter. Heat can move through gases and fluids by warm molecules rising and cooler molecules falling

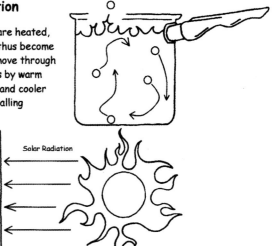

Solar Radiation

Conduction

Heat can move through an object by molecules warming, becoming active and bumping into and subsequently warming other molecules

The natural processes of convection and conduction spread heat through all kinds of materials.

Congratulations, you are now well on the way to becoming a fully independent, self-guiding solar thermal designer. As can be seen in this section, there is a lot of science going on 'behind the rays' in regard to making good use of sunlight to provide for our day-to-day needs. While some of it may at first seem like rocket science, it is really pretty easy to grasp most of the concepts that have been discussed. From earth science to physical science (and with a bit of common sense mixed in) we have explored the basic principles that can assist us in finding our way along the path to a better way of doing things. That stated, this is not the end, rather, it's just the beginning.

By finding natural processes that support the theories presented in the three laws of thermodynamics, we can reason that there is a seemingly endless transfer of energy that occurs all around the universe. The question for us then is, "how can we benefit in our day to day lives by having an understanding of this knowledge?"

Perhaps the answer lies ahead, in the remainder of this book and an exploration of solid objects that we can look at, pick up, hold and manipulate. For solar thermal design to become more than just a passing curiosity that people read about, talk about, think about and (some of us) write about, somebody needs to make some devices that work well enough to inspire others to do the same. In the next section, *The Nature of Materials*, I will discuss some physical characteristics of the materials that are commonly used to make devices that capture the heat found in sunlight. There is a little more science-type stuff to get through, but most of the way forward from here steps back into familiar territory and is centered on common objects that many of us work with on a regular basis.

The Nature of Materials

Chapter Six

Materials

Everything is made from something and solar thermal devices are no exception. All sorts of materials, from concrete to plastic, have been used to create commercially successful devices for the collection of solar heat. Knowledge of what has been used before, as well as a general understanding of how a variety of materials react when placed in contact with a heat source, allows solar designers to make the right choice for materials that they include in their devices.

This chapter is broken into a series of 'facts about' segments and as with most everything in this book, my preference is to keep the information as useful to the average solar heat dabbler as possible, while also remaining useable to those that have more experience. In this discussion of materials, I'm going to start with the thermal storage properties for a range of materials suitable for use as a heat storage medium, will then move to explaining insulation materials and will wrap up the chapter with an exploration of the role of color and surface coatings in relation to the overall effectiveness of a solar thermal device.

The majority of this chapter focuses on basic materials that are available to everyone, everywhere but just because they are common does not mean that these materials aren't effective; indeed they include one of the finest – water. The materials discussed will provide well-designed, solar heat collecting units with what could be considered up to 100% collection efficiency from naturally available, earth-bound materials and easily manufactured products. The advantage of using other, high tech materials could therefore be said to offer possibilities beyond that of a natural material; perhaps conceptually offering efficiencies greater than 100%. Such lofty goals can be reserved for the truly dedicated; those that wish to pursue the ultimate in everything that they do. The rest of us will do just fine working with what was made here on Earth, through natural processes over the past few billion years along with a bit of low-tech human intervention in more recent times.

Facts about:

Thermal Storage

In order to even out the daily variances caused by outdoor air temperature and to allow us to control the solar heating phenomenon in a way that can make the heat available to us after the Sun has gone down, a well-designed solar device may need to have some form of thermal storage. In certain situations there is no point in setting up systems that are able to collect lots of heat if there has been no provision for storage designed into the device. Interestingly, some commonly available materials are able to assist us very well in this endeavor. The ideal material for this purpose would be one that could readily accept all the heat that was delivered to it, was safe and easy to work with and was also capable of releasing stored heat at a controllable or regulated rate. A look down the chart on the next page will show some important properties in relation to the materials that could be easily integrated into a solar heat collector. The explanation as to what each property means is as follows:

Thermal Conductivity describes any material's ability to transmit heat throughout a specific volume of itself and is measured in watts per meter kelvin. The formula reads: W/mK, where a watt is the unit of energy, a metre is the unit of distance and kelvin is the unit of temperature measurement. To simplify the meaning of the equation think of it as:

"the energy transferred through a square meter of material in one second, when a temperature variance of 1 kelvin per metre is applied to the material".

It is also good for designers to know that thermal conductivity changes with temperature; for most materials it decreases slightly as their temperature rises.

When reading the chart on the next page, the higher the thermal conductivity number for a given material is, the easier it is for the material to conduct heat. **If we want to hold or contain heat, we need a low thermal conductivity rating.**

Materials

	Thermal Conductivity W m/K	Density Kg/m³	Thermal Diffusivity
Air	0.025	1.29	.019
Water	0.6	1000	.00014
Mineral Oil	0.15	920	.00009
Aluminum	237	2700	.0997
Copper	390	8960	.1130
Stainless Steel	16	7900	.0040
Concrete	1.28	2200	.0006
Marble	3	2700	.0012
Sand (dry)	0.35	1600	.0007
Hardwood	0.4	780	.002
Red Brick	0.6	1922	.04

The information contained in this table originated from a variety of sources, therefore the above figures are an accurate indication for each material, but not a definitive statement. Temperature variances and actual material composition can vary, altering the figures slightly.

Density tells us the weight of a material in relation to a particular volume of it. In the chart, the density is shown using kilograms per cubic metre. This factor can be important for two reasons – a high-density material will often hold greater quantities of heat than a low-density material and, especially if a device will need to be transported or intermittently moved, we can decide if the weight of the material is suitable for what is being built.

Thermal Diffusivity shows us the speed of temperature change in a material when it is exposed to a temperature change in its environment. Thermal diffusivity is therefore a measure of heat flowing freely through a material. It is defined as the thermal conductivity divided by the volumetric heat capacity (not shown here). **If we want material to either take in or dump heat quickly, we go for those with a higher thermal diffusivity number.**

For those interested in exploring the heat retaining/diffusing/releasing properties of materials further, two other factors, specific heat capacity and volumetric heat capacity, can also be added to the mix.

Water

As can be deduced from the chart, water is one of the finest heat storage mediums available. With a low thermal conductivity, water will slowly disperse gained heat throughout its mass. Its very low thermal diffusivity number means that it will give up its heat slowly and with a density that falls somewhere in the lower 'middle of the road' for the materials shown, it certainly isn't the heaviest storage material to work with. Oh yes, let us also not forget the biggest advantage of water when choosing it as a heat storage material – it typically doesn't cost anything.

Water does have a nasty physical property (well, to those of us trying to contain it) in that it expands when it freezes. If water is used for thermal storage in a location prone to freezing temperatures, protection must be in place to ensure that either the water doesn't freeze or that there is ample room for the system to expand if it does. The addition of automotive antifreeze to water is sometimes a way to avoid this problem. Another is to have the collecting system within a warmed facility and under a skylight.

Application: In what kind of situations might water be used for thermal storage?

Water is brilliant for use in containers located on the inside of large solar gain windows or other glass areas. The sunlight coming through the glass will warm the container full of water and after the Sun goes down the water will slowly radiate its contained heat to the room. The radiated warming effect in this type of system is almost subliminal, but it is definitely there.

Water is an excellent choice in isolated gain systems, specifically in a space heating application. Many old houses have heating systems that use fluid radiators, which are heated by a central boiler. After spending time living in such a house one can only wonder why electric heat ever took over; fluid radiators work very well and produce a pleasant feeling heat. Such radiators will continue to deliver heat long after the boiler fire goes out because of the warmth stored in the fluid and the same effect can be achieved when using a solar heat collector in place of the wood or fossil fuel-burning boiler.

Although it doesn't involve the Sun, one of my favorite applications for water is in a refrigerator. Inside of a partly filled refrigerator, any space not occupied by foodstuffs can be filled with water containers to stabilize the internal air temperature and reduce the overall running time, especially after the door has been opened.

Materials

Concrete

Concrete has got to be my number-two favorite choice for heat storage. Relating it to water, concrete's thermal conductivity and density are both double, which means that although it is more dense, heat is able to move through it quickly. Its thermal diffusivity is three times as high as water, so it will give its heat up far quicker.

Considering that it's heavy and seems to easily give up stored heat, why is concrete even on the list of choices? Compared to many of the other materials on the chart it is still a star performer on the heat storage playing field. It is also relatively inexpensive … not free, but cheap. Finally, it has a characteristic that most other materials don't; it is easily moulded to almost any shape. Wherever thermal storage is needed concrete can be fashioned to fit and it can also be easily colored to either match an existing paint scheme or be blackened to become more efficient at collecting solar heat.

Be aware that the factors shown in the chart for concrete will vary depending upon the nature of the aggregates used in its making. Thermal properties for concrete made in some areas may be superior to these results; others may not be as good.

Application: In what kind of situations might concrete be used for thermal storage?

Buildings, of course; masonry structures are among the most temperature-stable of all buildings and concrete floors can be turned into massive heat stores. If the sunlight coming in through a widow is allowed to shine directly onto a darkened, concrete slab floor, the heat will then be absorbed and spread by conduction throughout its mass. As with water storage units, once the Sun goes down the concrete will release the trapped heat into the surrounding room.

In an isolated gain system, concrete can be placed so that it can absorb heat from water or other fluids that are directed to flow over it or warm air that is blown across the concrete surface by fans or convective processes. Transfer ducts can be cast into the concrete that will allow the air or fluid to actually circulate within the concrete mass and extraction of the stored heat can be as simple as redirecting or reversing the air or fluid flow.

Combinations

Devices that use a combination of materials in their storage systems can offer the best use of products that are commonly available. Think of a situation that could use stainless steel, which is hopeless at storing heat but fantastic at conducting it, in combination with our favorite storage medium, water. Due to its thermal conductivity rating a stainless steel tank will easily pass collected solar heat on to water that is contained within. After the influence of the Sun is gone, the water will then release the heat at its own, much slower rate, while the stainless steel tank ensures efficient conduction of that heat into the surrounding environment. Without the water, the stainless steel is a poor choice; in combination with water we have a team that performs better than either material on its own.

Shall we venture to find another combo that could be useful? Have you looked at the numbers on copper? It is like stainless steel with a turbocharger attached; it will take heat in so easily but on the downside it will also release it just as easily. By using copper vessels as containment for mineral oil, a top performing storage system can be designed. Copper brings the heat into the system and the mineral oil, with its extremely low diffusivity, slowly releases it when the heat is needed … long after the Sun is gone.

Copper and water are another championship team although over time a chemical interaction between copper and water will form pinholes in the copper containment vessel. This takes years to happen but is certainly something worth knowing if you currently aren't aware of the problem.

Application: In what kind of situations might these combinations be used for thermal storage?

Most any time that collection and containment are remote from one another, such as in an isolated gain system, combinations of material are required. Picking the right combination of materials can make the difference between isolated gain systems that work well and those that lose much of the heat that they collect.

High Tech

Heat storage mediums that allow solar thermal devices to operate over an extended time period can be, as we have seen, very simple – water, metal and stone. In the pursuit of excellence some designers have gone for more complex, high tech options.

The commonly used refrigerant chemical Freon has been joined in recent years by engineered phase-change materials such as Glauber's salt (sodium sulfate decahydrate) and calcium chloride hexahydrate. Phase change materials have the ability to use chemical bonds to store heat as they range from a solid to a gel to a liquid and even a gas as their temperature increases. As they cool, the process works in reverse, chemically releasing heat as they return to their starting state, as shown in the diagram below.

While they are far more expensive, and in many areas more difficult to locate than water or other natural materials, their performance as heat stores can exponentially exceed that of natural materials. With products such as these becoming more commonly available, the reality is such that when a solar-powered, heat collecting system requires choices to be made about a heat storage medium, the size of the budget is just about the only constraint a designer will face.

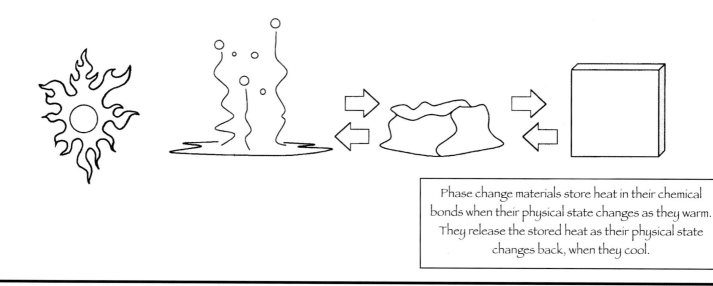

Phase change materials store heat in their chemical bonds when their physical state changes as they warm. They release the stored heat as their physical state changes back, when they cool.

Facts about:

Insulators

Insulating materials will help us to keep collected heat where we want it. When talking with others about the subject of insulation, many people will say that air is an efficient insulator, which it is … if it can be kept from circulating. The problem with using a cavity space full of air as insulation is that it will form convection currents, minimizing its usefulness by dispersing heat until equilibrium between the space and the colder external temperature is achieved. The most effective insulating materials combine low inherent heat transfer properties with the ability to hold air in such a way that it cannot circulate.

While the effectiveness of insulating materials can be judged by the same three factors as collecting materials – thermal conductivity, density and thermal diffusivity – we can simplify the selection criterion by focusing on just two factors: R-value and U-value.

R-value

R-value is a measure of a material's resistance to heat transmission and is an imperial measure calculation. The formula is $hr \cdot ft^2 \cdot °F / Btu$, which means:

"energy conducted, times inches of thickness, per hour of time, per square foot of area, per Fahrenheit degree of temperature difference, between the two sides of the material". Whew!

In countries that use metric measurement, which these days is nearly the whole world except for one really high-profile country, the R unit is called an RSI which stands for 'resistance to heat transmission using the International Standard Units of Measure' (which are metric units). The metric calculation is m^2K / W and that that means;

"the square metres of a material multiplied by degrees kelvin divided per units of wattage".

Okay, let's find out what these calculations mean and how all this works.

Materials

R-values provided by manufacturers for their products are fairly consistent and certainly reliable in commercial applications. The more experimentally prone readers can replicate the procedure that manufacturers use to determine a material's R-value and may find it fun to investigate things that they have lying around. Perhaps a pile of old, woollen sweaters or stacks of newspapers that just haven't quite made it to the recyclers depot, can be tested to see how well they rate against commercially available goods.

The testing procedure involves placing a heat source, like an incandescent light bulb, inside of a box made from the material to be measured. Because R-value calculations show a material's ability to inhibit heat flow, the temperature of the light bulb, and the temperature outside of the box are equally critical factors to consider. By using a probe-type thermometer to measure the temperature on the inside of the box, and a standard thermometer to measure the temperature on the outside of the box, a temperature variance can be determined and an R-value calculated.

With the light turned on, the temperature inside of the box increases and heat will be conducted through the material until, at some point, the rate of heat that is being conducted out of the box will equal the rate of energy being emitted by the light bulb. When this happens the flow of heat will be constant, providing a constant state of thermal transfer and the temperature inside the box will stop rising. When both the inside and outside thermometers maintain a steady reading, we can measure the R-value of the material (as determined from the inner and outer temperature difference) with the following equation:

R-value = $A(T_H - T_C)$/power of the heater

A is the area of the box in square feet (a one square foot box is recommended to simplify the calculation);

$T_H - T_C$ is the temperature difference between the hotter, internal temperature and the cooler external temperature;

Power of the heater is the wattage rating of light bulb.

Because this is an R-value calculation, all units must be imperial. The power of the heater therefore needs to be calculated in Btu rather than metric watts. The factor for converting watts to Btu is: number of watts x 3.41 = Btu.

If you're in a metric country, RSI can be determined from the final R-value by simply dividing the R-value by 5.6. Good luck to those keen on trying this experiment.

Common Insulators

While there is an enormous range of materials that have been used in insulating applications, the reality of economics and ease of use will likely bring solar designers back to the two most common insulating materials in the building world, fibreglass batts and polystyrene boards; both are available in a range of thicknesses and varying R (or RSI) values.

Fibreglass in the form of home insulation batts is so easy to work with. The batts are simple to cut to size, available at every hardware shop I've ever been to and inexpensive. The little fibres of spun glass do seem to always get up my sleeves or down my back, but as long as an itchy arm (or body) isn't a bother, fibreglass batts will probably be at the top of every solar designer's list of insulation materials. The itchy bits … well that's just a small trade-off when working with such a universally applicable material.

Polystyrene, commonly known as Styrofoam, comes in two types – expanded cell and extruded. Expanded cell is easily distinguished from extruded; expanded cell is the stuff that crumbles into a million little balls when it is sawn or broken. It has a good R-value and works well in devices that can be designed to encase it within a sandwich of more robust materials, like plywood or solid timber. It should never be used in applications where it is openly exposed, as it is damaged easily, degrades quickly from sunlight and can absorb moisture, which causes it to loose its insulative ability and even rot.

Extruded polystyrene is the 'Formula One racing car' of insulating materials. It has a comparably high R-value in relation to other insulations for a given thickness, is very strong and cuts cleanly with a just a knife. It too is best applied as part of a material sandwich between layers of plywood or timber as it is also damaged by sunlight and handling but it will not absorb moisture. Though it can be difficult to find in some areas, a source for extruded polystyrene should definitely be found. The only reason I rate fibreglass at the top of the insulation list is because everyone can obtain it easily … my number one pick personally is extruded polystyrene.

Materials

The chart on this page lists both the R-value and RSI-value per inch of the most commonly found insulating materials. For metric users, please consider the inch to be relating to a 25mm thickness of the material. Note that at the bottom of the chart are ratings for some materials not commonly associated with insulation but used in the construction of solar heat collectors. We'll see why they're listed, on the next page. Also notice that air, at top of the list, shows a similar value for a space ranging from one to four inches. Also, in items like plywood, regional differences in timbers and glues may alter their rating.

If insulative value conversions are ever necessary, the conversion factor to go from R to RSI is:

R x .1761 = RSI

and to go from RSI to R:

RSI divided by .1761 = R

Material	R-value per inch	RSI per inch
Air	A 1 to 4-inch space = .9	0.16
Mineral Wool	2.5	0.44
Vermiculite	2.5 - 3	0.44 – 0.53
Cotton	3	0.53
Cellulose (straw)	3.5	0.62
Fiberglass	3.5	0.62
Expanded Polystyrene	4	0.7
Extruded Polystyrene	5	0.88
Polyurethane	6	1.06
Concrete	0.08	0.01
Brick	0.20	0.04
1/2" Gypsum Board	0.45	0.08
8" Concrete Block	0.58	0.1
Soft Woods	1.25	0.22
Plywood	1.25	0.22

R-value Composites

The chart on the previous page shows listings for a few construction materials because accurately determining a containment mechanisms' R-value involves adding the values of all materials involved in its outer skin. Insulating enhancement is primarily achieved by either increasing the quantity of a material that is being used or by upgrading to a higher quality material, say trading out fibreglass batts for extruded polystyrene boards. The only limit to improving the overall R-value of a containment device is space available in the cavity between the devices inner containment area and its outer skin.

In addition to the R-values for the various components of an assembly, there is also a surface film of stagnant air that is caused by the molecular interaction between air and solid surfaces. While it varies slightly between indoor and outdoor surfaces, and again between different materials, we can safely estimate an interior surface film R-value of 0.6 and an exterior surface film of 0.2. The exterior is lower because it is subject to a much wider range of environmental factors: wind, rain, air temperature fluctuation, etc.

The diagram on this page shows an example listing the R-value components of a typical well insulated, timber-framed exterior wall with a 1/2-inch plywood outer lining and a 1/2-inch drywall inner lining. Notice how each components' R-value contributes to the overall rating of the wall.

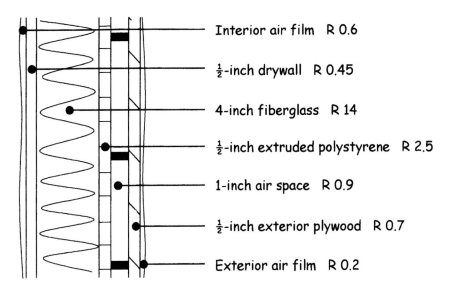

Interior air film   R 0.6

½-inch drywall   R 0.45

4-inch fiberglass   R 14

½-inch extruded polystyrene   R 2.5

1-inch air space   R 0.9

½-inch exterior plywood   R 0.7

Exterior air film   R 0.2

Cumulative R-value total = 19.35

Materials

U-value – The Missing Link?

The most common reference used in insulating decisions is the R-value. The higher the R-value of a material, the better it is at resisting heat loss (or heat gain). While high R-values can be achieved by stacking inches upon inches of insulative materials, the real effect of insulation, a value known as the U-value, diminishes with each additional thickness of material. Where R-value indicates a material's *resistance* to heat flow, U-value measures the amount of heat that will actually *flow through* a material, which results in heat leaking from warm areas to cool ones. In the imperial measurement system, the U-value is the number of Btus of energy passing through a square foot of the material in an hour, for every degree Fahrenheit of temperature difference across the material: $Btu/ft^2hr°F$. In metric units, it's calculated in watts per square meter per degree kelvin: W/m^2K.

Do these formulas sound more than a bit like thermal conductivity? It's actually not too far removed. The metric calculation for U value is W/m^2K; very similar indeed to the calculation for thermal conductivity, W/mK. As can be noted from the above explanation however, U-values deal with the transmission of heat *through* a certain square footage or metreage of a material while thermal conductivity measures the spread of heat *within* the material.

Of importance to designers of heat containment systems, is that U-value has an inverse relationship to R-value; as the R goes up the U goes down. We are looking for lower U-values to signify better insulation as a result of lower levels of heat transmission. To calculate the U-value for any material, it is easiest to start with the R-value. Simply divide 1 by the R-value; that is all inverse division really amounts to. An R 20 containment system can easily be calculated to U .05 through dividing 1 by 20.

What does all this mean? The biggest insulative impact will be made with a device's primary insulating material. As additional materials or material thickness is added the insulative value does increase, however the percentage of the increase steadily reduces with each additional R-value. U-value calculations let designers determine a point of diminishing return, where the additional expense of more insulation does not equal a proportional increase in overall insulating ability. Have a look at the simple table on the following page to see how this works.

Insulative Material	Component R-value	Total R-value	System U-value	Performance Improvement over Baseline
1-inch plywood box	1.25			
Outer surface air film	0.2			
Inner surface air film	0.6	2.05	0.48	Base line
Outer layer of 2-inch extruded polystyrene	10			
New exterior 1-inch ply layer	1.25	13.3	0.075	84.5%
Another outer layer of 2 inches extruded polystyrene	10			
A new exterior 1-inch ply layer	1.25	24.55	0.041	91.5%
2 more inches of extruded polystyrene	10			
Another new 1-inch ply exterior skin	1.25	35.8	0.028	94%

A plywood box makes large leaps in its R-value as nine different insulative elements are eventually figured into its design. Note however that it is only the first alteration, adding an outer box and filling the resulting cavity with polystyrene, that really makes a significant difference to the devices' true insulative value, the U-value – its ability to stop the transmission of heat from within the box.

Materials

When setting up naturally influenced systems that will benefit our lives, we must be able to control where the heating and cooling effects resulting from radiation, conduction and convection take place. As we know, for a material to transfer heat effectively it must be dense like concrete, stone and water. In general terms, the more density that a material's molecular structure has, then the more effective it is as a medium for heat transfer. Alternatively for a material to be a good insulator it must have a light molecular structure. If a material can trap air or other gases within its molecular structure, it is likely to be a better insulator than those that are more dense.

The diagram on the following page illustrates the concepts that we have learned in this discussion of insulation along with ideas that we discussed back in the chapter on Thermodynamics. By using the right combinations of materials, so that heating effects are allowed to happen in combination with materials that have an insulative value, we can keep the heat where we want it and improve the efficiency of the solar thermal devices that we make.

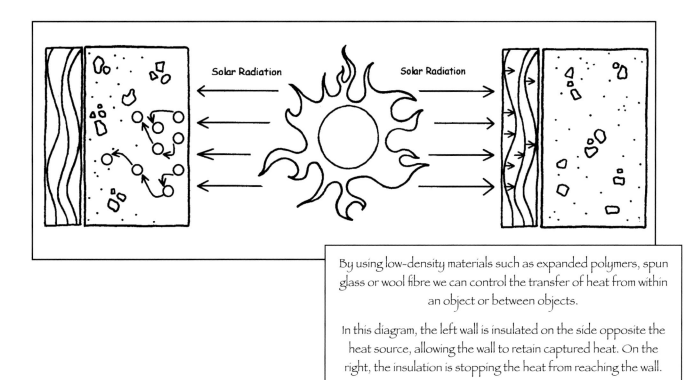

By using low-density materials such as expanded polymers, spun glass or wool fibre we can control the transfer of heat from within an object or between objects.

In this diagram, the left wall is insulated on the side opposite the heat source, allowing the wall to retain captured heat. On the right, the insulation is stopping the heat from reaching the wall.

Materials

Facts about:

The Role of Color

The final factor to consider when choosing materials for our devices is their color. As this gets a little scientific, this fact sheet should have probably appeared back in *The Science Behind the Sunlight*, but I liked the way it fit in here with the materials.

In the purest form of solar thermal design, the only choice for maximum heat collection from sunlight is flat black but in order to better understand why colors such as black can absorb more heat from sunlight than others, we should start by taking a look at the very nature of light.

Light

The light that can be seen coming from the Sun, a light bulb, a candle or any other source is just the visible portion of a broad spectrum of what are known as electromagnetic radiation – remember that term from back in Chapter 4? We are exploring it farther here because there is actually far more to the radiation in light than we can detect with our eyes. The radiation is made up of waves that vibrate at different frequencies; a frequency being how many times the wave oscillates (to simplify this, imagine it going up and down) in a second. Every electromagnetic wave exhibits a unique frequency and a distance between the peaks of the waves, known as a wavelength, that is associated with its frequency.

For instance, the radiation that we see as the color red has a wave frequency of 428,570 billion times per second. This means that when looking at red light, a person's eye receives 428,570 billion waves every second. The wavelength of the red light is just 700 nanometres long, which means that the space between wave peaks spans only 7 ten millionths of a meter. Since radiation always travels at the constant speed of light, which is 300 million meters per second (669.6 million miles per hour) the wavelength is a convenient measure for the different frequencies.

Materials

The fastest vibrators are gamma rays, followed in sequence of slower frequency by x-rays and ultraviolet rays; we cannot see any of these with our eyes. Moving to still slower frequencies, we will then find visible light in the blue spectrum, visible light in the red spectrum and all other colors that fall between them such as green, yellow and orange. Finally, really starting to slow down in frequency we come to more wavelengths that we cannot see; infrared rays, microwaves and finally, the slowest of all, radio waves.

The science of light is very complicated. Consider the nature of a wave and how it might interact with another wave when the two are travelling along the same path, side by side. If two wave peaks were to collide, the impact could crush the peaks and cause a shorter wave to result. If strong enough, the collision could actually cancel out the two peaks or perhaps the waves will meet each other when both are on a rise; in this case they may supplement one another and create a higher peak. This phenomenon is also true of the interaction of wave troughs. They too can decrease, cancel or supplement depending upon the nature of their interference with one another. The phenomenon is properly known by the term **interference pattern** but as it happens to turn out, waves and interference patterns are not all that there is to know about the science of light.

Type of Light	Wavelengths
Radio Waves	Greater than 30cm
Microwaves	30cm – 1mm
Infrared	1mm – 780nm
Red	779nm – 622nm
Orange	**621nm – 597nm**
Yellow	**596nm – 577nm**
Green	**576nm – 492nm**
Blue	**491nm – 455nm**
Violet	**454nm – 390nm**
Ultraviolet	389nm – 10nm
X-rays	10nm – 0.01nm
Gamma Rays	Less than 0.01nm

This table shows the length of the light waves for the full, known range of electromagnetic radiations that create light. The abbreviation nm stands for nanometer – one ten millionth of a metre.

Only those shown in bold are visible to the human eye.

Albert Einstein challenged the Wave Theory of Light in a 1905 paper, when trying to interpret an experiment in light that had been performed by Max Planck. In this paper Einstein proposed that even though light travels in waves it must also contain a specific amount of energy, a packet of energy, that could be further interpreted as a particle of light. He called these particles of light **photons** and like with so many of Einstein's proposals, the theory was proven to be correct through subsequent experimentation. So what does this finding do to our understanding of light? Does it help it, hinder it or change it?

Well, it actually augments it. Light is *both* a wave and a particle and as Einstein explained of his findings, the nature of light involves a fusion of both the wave and particle descriptions. Light is both and it is either, depending on how it is acting at any given moment. There are no 'solid' comparisons; light is the only physical thing so far discovered to work like this, so for many of us the concept will be hard to understand because we cannot relate it to anything that we can touch.

The accepted term within the science of quantum mechanics for the phenomenon is **wave-particle duality** and sorry, I have no diagrams for this one – it is one of those times when we have to put our faith in the hands of people that work in such matters. This is just the way it is.

Materials

All the Colors in the Rainbow

The heat from solar radiation is absorbed and/or reflected at different rates by different colors. White materials reflect most of the light they receive and therefore absorb very little radiation. Yellow materials absorb blue light, blue materials absorb red light, purple materials absorb green light. The wave frequencies that they reflect therefore make up the color that we see. Black materials absorb all the light frequencies that they receive, which is why they are so well suited to use in solar thermal devices.

This information can be put to good use when designing solar heat collectors. The following list is in order of effectiveness with the best absorbers are at the beginning of the list:

Flat Black, Dark Grey, Dark Brown, Dark Green, Brown, Rust, Red, Grey, Orange, Yellow, Blue

Research has indicated that when dealing with colors, other than black, that are of the same tone on a grayscale, browns and greens are the most effective absorbers followed by reds and finally blues.

Grayscale – What is grayscale? If a black and white photo was taken of a range of colors, the photographic image that will result will show those colors in a shade of gray. Lined up side by side, from lightest to darkest, the black and white images of the colors create a grayscale.

Facts about:

Heat Absorption & Emission

When sunlight strikes the surface of a given material, basically two things can happen – the radiated wavelengths can be absorbed and retained or they can be re-emitted. If the wavelengths are fully re-emitted, the material will be seen as white, which is the color produced by the combined frequencies of all electromagnetic radiations. If they are fully retained, the material will be seen as black. From this understanding we can reason that red, blue, green, orange, yellow and every other color of the rainbow are produced when a material absorbs some frequencies and emits some others. We see a material as having a color because whatever we are looking at is re-emitting those frequencies that create that color back into the environment, while absorbing all other frequencies of radiation from the visible spectrum.

When a material absorbs light, it is also absorbing all the energy present in the waves and photons that are striking its surface. As we know from the section on thermodynamics and the laws of energy conservation, the light's energy cannot just disappear upon being absorbed – it must be converted into another kind of energy. The most notable conversion that takes place transforms the electromagnetic radiation into kinetic energy within the atoms that make up the absorbing material. If the energy is retained, it will raise the temperature of the material as a result of greater molecular activity – kinetic energy. If it is unable to be retained, the object will release the absorbed energy by re-emitting the lower intensity, longer wavelength of infrared radiation, which is essentially heat. This transformation of light into heat is the key to understanding the process because it accounts for the law of conservation of energy in this chain of events. While color has no effect on conduction or convection, it does in this way affect radiation.

The key point to remember in all this is that darker objects feel warm when in contact with sunlight because they are better at emitting heat due to the fact that they are *also* better at absorbing the energy in light to begin with.

Materials

Planck's Law deals in blackbody radiation, a blackbody being an object that absorbs all light that falls on it. A blackbody will not *reflect* light, but it is still capable of emitting energy in the infrared range after first absorbing all of the light that it has received upon it. With a basic understanding of this concept of absorption followed by emission, Planck's Law supports the idea of using collecting surfaces that have a greater ability to collect heat and then pass it to a storage medium than those that immediately reflect all or even some of the incoming light.

The fact that everything at any temperature emits light might seem a bit obscure. When looking at everyday objects we don't usually think of their light emitting properties, but then much of the light coming from common objects is far into the infrared; we can't see it, but it is present as evidenced by the object's temperature. Planck's Law states that the wavelength of a radiation emitted *from* a surface is proportional to the temperature *of* the surface. A material that is classified as a perfect emitter will therefore release the same amount of energy in the form of infrared waves that it had received from every wavelength of light it was capable of absorbing. Would such an efficient emitter be useful to suggest in a solar thermal design, even if it were a powerful collector?

No material has been found that can absorb all incoming radiation, however it is largely accepted that the best color for collecting solar heat is flat (non-reflecting) black. Further investigation will reveal however that all flat blacks are not created equal. We will discuss selective surface coatings ibeginning on just the next page, and this discussion on emissivity is a great lead in to that area of discussion. We need to understand that flat black is not just flat black; we need to know what the material that looks flat black actually is in order to know if it is the most suitable flat black for our purposes.

Most black paints work well, black dye in water even works well but nothing is even close to graphite when it comes to absorbing ability. Graphite absorbs all but about 3% of the light waves directed at it, so it would seem to be the top choice for a collecting surface. On the downside, graphite is also a perfect emitter of radiation; at a particular temperature a graphite object will emit the maximum amount of energy possible for that temperature. This paradox can be found again and again in surface materials, the best collectors are also the best emitters. In something of an answer to my question above, the search for surface materials needs to be focused on finding those that strike a balance between effectively taking heat in and then if they cannot assist in holding that heat when the Sun goes down, using a suitable material in combination with them that can store the heat more effectively.

Selective Surfaces

The surface of a solar collector can be coated with materials that are known to produce a specific heat collecting/emitting characteristic. Any surfacing products that produce such a controlled result is known as a **selective surface**. As mentioned, the best collectors are also the best emitters, if we reduce a material's ability to emit then we also will typically reduce its ability to absorb radiation from the Sun. Are we playing a game that we cannot win? Not exactly, the trick lies in creating a surface that has multiple characteristics and this will involve layering the right materials in an order so they can do what we want them to do.

A well-documented selective surface has been created from stainless steel that has been layered over first with gold and then with silicon. The outer silicon layer looks black to visible light but is essentially transparent to infrared light. Because of this, when on its own the silicon would rapidly spill off any heat that it absorbed. By virtue of being layered over the gold coating, the overall surface emits like gold for infrared emission, which is very low at 10% for infrared wavelengths, but not before the heat has been carried through to the stainless steel where it is stored. This combination is therefore an excellent absorber of sunlight as well as a poor emitter of heat (light in the range of the infrared wavelengths). If someone were to use this silicon/gold surface over a stainless steel fluid container, they wouldn't want it filled with water unless trying to produce industrial grade steam; the combination can apparently rise to a temperature of 375°C Awesome performance ... but at a high cost in dollar terms to create.

Does the average designer with limited financial backing have a selective surface option that will work well? Black bar-b-que paint over copper is a good choice on the low-cost end of the available options. The paint is engineered to withstand high temperatures and the enamel base layer that carries the black pigment will function as a barrier against coppers' thermal conductance and diffusivity issues. The paint also assists the copper in collection by providing a dark, non-reflecting surface that copper doesn't have on its own. Care needs to be used with this combination as well; it can exceed the boiling point of water, just not quite as far as the gold and silicon option.

There is a wide range of surface materials and application methods to choose from. The material itself dictates some application methods; others are simply convenient ideas that produce an acceptable result. As usual, some are high tech and some are low tech.

Oxide coatings are the epitome of low tech; just leave a metal surface exposed to the elements of nature and it will oxidize. Metals such as copper and iron that were used in solar collectors by Mouchot, Ericsson and others underwent a natural oxidation process. For some materials this was discovered to have a desirable effect on absorption characteristics when compared to the material's natural or new state. The problem with this method is that any oxidation processes occurring naturally, are naturally not under our control. The changes in absorption that result from natural oxidization happen at their own pace and can also change the characteristic of the material's ability to retain heat. In some materials, oxidation can cause a slight reduction in the efficiency of the collector.

Chemical coatings such as enamel, lacquer or epoxy paints are the working class choice of surface finishes. While usually applied by either spraying or brushing the coating onto a collector's surface, there are some options, like those applied with the process known as powder coating, which use electricity to assist their application. Chemical coatings do not alter the absorbing or emitting properties of the underlying material; rather they enhance its ability to absorb solar radiation by layering on a material with the more desirable surface characteristic of better absorption and/or lower emission.

Despite their relatively low cost and ease of application, some chemical coatings can literally be burned off by the temperatures reached inside of solar collectors. For example, standard black enamel paint applied to a metal plate is likely to melt or burn off when the combination is repeatedly brought to high temperatures. The thickness of a chemical coating can also make a difference to how well it works. Control is again a factor to consider as the thickness of the surface coating can alter the collectors' ability to both absorb solar radiation and subsequently emit heat.

Electroplated coatings are the most widely used selective surfaces in commercially produced solar collectors. This type of coating can only be applied to metal collectors as it uses electrical current to fuse metals such as brass and chrome to a base metal like copper or steel. Traditional electroplating technology is available to anyone located near a major city; there is no mystery to the process and it is not very expensive to have done. Although a relatively rare process now-a-days, black chrome was once a common coating used for custom car parts. It has experienced new favor in the manufacture of some commercial solar collectors and in cities where shops can be found that produce black chrome, designers of solar heated devices have good fortune smiling on them. Chrome is a relatively stable coating and does not degrade even with regular exposure to industrial temperatures as high as 392°F (200°C). High humidity can cause a very slow oxidation process to take place with chrome, but this is not really much of a problem if the system is reasonably maintained. On the downside, chrome coatings are reflective so some wavelengths of light are lost.

Vapor deposited coatings are the ultimate in surface coating technology. They are not so much used for directly coating a collecting surface, as for the coating of reflecting surfaces. Aluminizing and silvering of mirrors has been common practice for over a hundred years and some methods produce a reflectivity index of greater than 97 percent. The process is quite expensive as it is carried out in a vacuum chamber and to be honest, I doubt many facilities could be found that were capable of handling something as large as a solar reflector. For those working with nearly unlimited budgets, have a look around to see what services can be provided by facilities that work with vapor deposited coatings, the more light that is reflected onto a collector the more heat it is capable of storing.

Materials

Facts about:

Glass

Glass gets a page of its own because it ranks fairly high up on the range of materials that all solar thermal designers will need to use. I have placed the page on glass last because glass provides solar heated devices with their last opportunity to capture heat from solar radiation. The key factor that makes glass so important to our devices is that while sunlight can easily pass through its clear surface, the lower energy wavelengths found in the infrared range have a much more difficult time with this task; they become trapped behind a pane of glass and this can dramatically boost the temperature of a solar collector.

As mentioned in the discussion on blackbody absorption and emission, balance is often the key in designing a solar thermal device. When energy from the increased molecular activity warms the collecting body to a point where it cannot retain any more heat and the body begins to emit infrared, a glass-covered box can then capture the infrared waves being radiated from the collecting body. Once a thermal equilibrium is attained between the total system of materials at work in the collecting device and their surrounding environment, in other words when the U-value is exceeded, the device will have captured the total amount of heat that its design can retain.

All glass is not created equal but all glass seems to work well at retaining some amount of additional heat. Two layers does the job better than one, but like all insulative factors, the U-value of a third layer provides a diminished return. The practice of using two layers is known as double glazing. The world of high tech has made it to this practice as well, and an even greater result can be obtained if the dead air is removed and the space between the glass layers filled with argon gas. For a good result without the expense of commercially manufactured argon-filled units, just make sure the space between the two panes of glass is well sealed, has a desiccant pack added to remove any errant moisture and does not exceed one inch; convection currents cannot form in a space less than one inch and therefore the space of 'dead air' can insulate reasonably well.

Materials

In this section we have explored a wide range of properties that apply to the materials used in the construction devices that collect solar heat. Without these basic units of knowledge, the best a designer could hope for would be a continued range of experimentation with what has likely been produced before. With an understanding of the basic material properties contained in these past 31 pages along with the principles explored in the previous section, solar designers are free to lay awake for hours at night, annoying their bedmates as they design, redesign and ponder further designing possibilities in their minds. Poor Sabrina (my partner) she has demanded me out of the bed that many times as I ponder solutions and new ideas that I can no longer count them; I am also no longer offended by her wishes for a peaceful night's sleep. Poor me in that my mind just works that way … but I wouldn't really want it otherwise … I like my mind so I guess it's actually 'lucky me'.

In the next section, *Practical Applications*, we can finally get some ideas on how to get our hands dirty and utilize solar designing principles in the design of devices that we can use. The devices range from the relatively mundane to the eccentric, but all of them could be considered by anyone reading this book as a step along the way to provide themselves with a cheaper, possibly better aspect to some part of their lives while maybe, at the same time, driving in the thin end of the wedge that will help to further open the door to a world not so reliant on fossil fuels.

Come along, the adventure is just beginning.

Practical Applications

Chapter Seven

Application of the Knowledge

Learning for the sake of learning is, in my opinion, a worthwhile pursuit. I love to learn new things, even topics that have absolutely no way of ever finding themselves as part of my normal existence (provided they are at least interesting to me). However, learning with the intent to apply that learning is another level of existence altogether. As a management consultant and trainer, I am well aware that the practical application of new learning not only reinforces the learning but typically leads to new ideas and innovation as well.

What I usually expect to find in the final section of most solar-designing information books is a list of projects for inspired readers to go off and build, changing their lives forever to a friendly existence in concert with the Sun. I considered doing this as well but I don't think that there is much for me to contribute to the ever-growing list of regurgitated solar device projects. Since the info already discussed in the *Principles and Materials* sections will have everyone clued up as to what material combinations they can effectively use when designing solar heat collectors, I will choose to leave the device designing to the readers.

In place of a projects section, I thought I would provide a little more information as it relates to a few examples of solar designing projects by myself and others along with some narrative about them. Some of these, like my home, have been successfully completed and function well. Others, like a solar toilet, were successfully completed and whole-heartedly rejected by the rest of the family. Still others are just experiments, maybe providing a bit of design inspiration to others.

As has been demonstrated throughout the book so far, solar designing is not rocket science, just a conscious combining of the right materials in the right place to let the Sun provide us with a useful source of heat. It's up to us to be creative with how we use it.

Study One

Sun Tea

Cooking is a part of everyday life and around our house another part of everyday life is a cup of tea; or several cups of tea. While Sabrina has very little interest in drinking it cold (and from what I gather most other non-Americans agree with her) I love iced tea. Regardless of how you drink it, the Sun makes a beautiful cup of tea by just being left alone to do it.

This one is really just too easy. I put ten regular tea bags into a wine-makers demijohn and leave it in the Sun for a few hours (Sabrina is the wine maker if you were wondering). I find it takes a little longer to brew in the winter, less time in the summer.

For iced tea, it is ready straight from the demijohn; I just add ice, sugar or whatever else to suit my taste at the time, like fruit juice. For hot tea it goes from the demijohn to the kettle for a few addition minutes over the open flame … or in front of the solar concentrator (coming up).

Done.

Application of the Knowledge

Parabolic Cooker Experiments

Experimenting with parabolic reflectors is a great way to bring back the feelings of being a curious little kid doing something that your mother told you not to. Like so many youngsters, fire fascinated me. The magic of striking a match and watching it burn seemed to hold some locked up secret of the universe. Well, some time after getting in big trouble for playing with matches, I learned how to burn things by focusing the Sun's rays with an old magnifying glass that had belonged to my grandfather. For some reason this was even more remarkable to me than the matches. It was also quite likely one of the first scientific experiments I ever carried out and was absolutely my first recollection of harnessing the power in the Sun's light. I still have that magnifying glass but now that I'm a big kid, I also have something that, in my opinion, is even better: not one, but two parabolic reflectors.

The process of tinkering that led to these two reflectors was actually a lot of fun. The first reflector is completely homemade and is of marginal efficiency. Being constructed from cardboard and aluminium foil in a single evening, this has to be a classic recommendation for anyone that wants to have a play with heat generated from sunlight. The second, created from a satellite TV dish antennae that I happened across in a salvage yard, is a good example of how an unsuspected find can lead to innovation when combined with an existing knowledge base.

Collapsible Parabolic Reflector

Although I said I wasn't going to outline any projects, I have to do just one. This collapsible parabolic reflector is so neat to make that I can't help but provide a simple how-to, including the basic dimensions, materials and build instructions. I decided to make this device one evening after coming across a design for one in a book called *The Solar Boat Book* by Pat Rand Rose. Made from a cardboard box that we had lying around, I cut out all the pieces and spent an entire evening engrossed in the mostly mindless process of making it. Mine is a half-size version of what was outlined in the book and will not boil water, as the larger version is supposed to do. In about 20 minutes it will bring water to a point that is just steaming; hot enough to make a cup of tea, coffee or a bowl of instant soup, which was good enough for me.

The build is simple:

First, cut out 18 cardboard wedges. The wedges are 9 inches (225mm) long, 1/2–inch (15mm) wide across the small end and 3 ½–inches (90mm) wide across the big end. They have two tabs that protrude by 3/4–inch (20mm) and both are 1¼–inches (30mm) long; the tabs come out from the same side on all of the wedges. Placement on these tabs is not critical, just get close enough to what you see in the accompanying photo (next page, top left), which has the tabs placed at 1¼–inches (30mm) and 5¼ inches (130mm) from the big end of the wedge.

Lay the wedges side by side, with the tabbed sides overlapping the straight sides and mark on each wedge around where the tabs overlap. Using a knife or screwdriver, rip the core out of the cardboard within the area of overlap for both tabs on the straight side of each wedge.

With the core removed, the tabs will be able to be inserted into the un-tabbed sides of each wedge. Insert them, one wedge at a time and drive a toothpick through the wedge surface, the inserted tab and out the back surface for the tab that is closest to the big end so the wedges will stay connected. When all are pinned (toohpicked) you will have a cardboard disk that magically takes on the shape of a parabolic dish, with a hole about 2 inches (50mm) in diameter right in the center.

To secure the assembly and to use as a mounting surface, I had already cut out two plastic discs, each 4 inches (100mm) in diameter and with a 5/16–inch (8mm) hole drilled in the center. To complete my test assembly I sandwiched the dish between the plastic discs and clamped the discs tight with a bolt, washer and wingnut combination.

Disassemble the test-fitted dish and, to cover one side with standard aluminium foil, first spread a thin layer of white glue on the cardboard wedges. Place the foil shiny side up with as few bubbles or wrinkles as possible and leave it to dry at least overnight.

On my project, once the glue was dry and the excess foil trimmed off, I reassembled my reflector dish, sandwiched it between the plastic discs and secured the assembly to a standard camera tripod with the point of focus aimed at my kettle.

Application of the Knowledge

Parabolic Dish Antennae

I think that the antennae that I made this from is actually not meant to be removed from the homes they are attached to, actually belonging to the satellite TV company that services our area. When I came across it in a salvage yard, I just couldn't help myself wondering how much they wanted to sell it for. Expecting something in the hundreds, I was surprised when they said "twenty dollars".

"Mine! I'll take it", was my reply as I reached for my wallet.

I then asked about the legality of trading in such dish antennae's, to which the conversation went rather flat; something like "we don't leave much when we do a demo job". No need to push the issue smiling at my good luck I headed for home.

The antennae had a latex-like paint on it that was rather difficult to remove, even with an industrial paint stripping chemical. Patience pays off in such things as stripping paint and after a second day of chemical applications and careful removal of the lifting paint (so as to not scratch the metal surface) the face of the dish was down to bare metal. I realised that luck was really shining on me when, after lying in the garage exposed for several days and showing no signs of oxidation, I discovered that the dish was stainless steel. This meant that there was no need for clear coatings, which would have reduced its reflective ability.

With the paint stripped, I then had to polish the metal surface. Metal polishing is pretty straightforward work that basically involves sanding the surface with progressively finer grits. I started with a 240 grit paper to eliminate the fairly rough surface texture that apparently is meant to hold the paint, worked through a series of 320 to 600 grit to get a shine started and then finished sanding with a 1000 grit leading to a 1200. This really got the shine up and the final polish was achieved with a commercial, metal polishing compound.

Although this dish is only slightly larger than the collapsible one, the manufactured precision in the parabola and the smooth, reflective surface really make a difference. With exposure to the dish's focal point, wood will set to fire in a matter of seconds and water in the kettle will boil in about 15 minutes (see photo, opposite page). With a small cast iron pot, it can make a nice stew-for-two in a just few hours, without even needing to be repositioned as the Sun's position changes in the sky.

This satellite TV antenna polished up nicely to become a solar concentrator.

Application of the Knowledge

Solar Thermal Electricity

While stepping beyond the realm of the average home workshop, the generation of electricity as a result of solar generated heat is still being explored, expanded and, most importantly, used in commercial applications.

Solar troughs, as displayed in the pages about John Ericsson, are one favourable method. Power towers, displayed in the pages on William Adams are another. Dish-type concentrators like those pioneered by Mouchot and Eneas have continued to draw the attention of solar thermal designers, especially due the extremely high temperatures they are capable of producing.

During 2000, at its STAR Center in Tempe, Arizona, APS, Arizona's largest electric utility provider, was evaluating the performance of the Parabolic Dish/Stirling Engine solar power system shown in the photo on this page. This unit is capable of producing 25 kW of electricity by using the mirrors to focus sunlight onto a thermal receiver. The heat is used to run a Stirling heat engine, which drives an electric generator. Quite a step up from my little parabolics!

Photo courtesy of National Renewable Energy Lab
Image credit: Bill Timmerman

Solar Ovens

Trying to find applications that can be adapted to suit many people with different needs can be a challenge. Everyone needs to eat however, so one sure way to provide something for everyone is to look at solar ovens, which can be sized to suit families, couples or even made portable enough for individuals out camping.

Cooking food with Sun-power is an enlightening, almost liberating, experience. The first time I ever tried it involved some soup in a polystyrene take-home container from a restaurant. I took the soup with me to a meeting that I knew would run for a couple hours and placed it on the dashboard of my car before going in. When I came out a couple hours later ... hot – I mean *really* hot – soup was waiting for me. I was convinced by the effectiveness of such a simple procedure and intrigued at the same time. It just seemed so *free*.

Photo courtesy of National Renewable Energy Lab
Image credit: Tom Lawand

The ovens shown in this photo are form a community in Lesotho, South Africa. As can be seen here, one can cook food for large groups, the other is more suited to a family situation.

Application of the Knowledge

Solar ovens are a simple to make, well-tested device that go back to the 1767 and Horace de Saussure. They are simple to use and can be produced from materials as common (and cheap) as cardboard and plastic wrap or with any degree of greater materials sophistication that the designer chooses to integrate. My own is not featured here because of a rather unfortunate incident involving it and my bulldozer just before I began taking photos for this section of the book, so perhaps I should refrain from the 'simple to use' statement. Maybe there is a warning for all in that little story; no major landscaping is to be done in the vicinity of a solar oven!

The basic design of the most effective cookers involves an insulated box made waterproof on the inside, covered in a double glazing on one face and utilising one or more simple reflectors to bounce additional sunlight into the box. An oven that is sized properly, with adequate reflecting area, will not need to be turned as the Sun passes overhead and the device is especially suited to slow cook items like stews, crock pots and soups. As can be seen in the photo to the right, solar cookers can do a pretty good job of baking as well.

From stews and soups to baking cookies, solar ovens can provide for our cooking needs.

Photo courtesy of National Renewable Energy Lab
Image credit: Warren Gretz

The Solar Toilet

Try to convince your 'normal-lifestyle' partner, used to all of modern life's conveniences and on a steep learning curve in the alternative energy scene, that she and her children, also used to those conveniences but with no interest in learning about alternative lifestyle concepts, should do their private business while sitting on an insulated box with a window on it. Good luck is all I can offer to those keen enough to give this scenario a go. Unfortunately (maybe fortunately?) luck wasn't enough for me in this endeavor.

I first came across the solar toilet concept when digging through alternative lifestyle books in the Antarctic. The basic idea is a modified version of the solar oven; heavily insulated walls and a glass front combine to create a greenhouse effect that can cook food ... or faeces, depending on how you use it. This really is a sensible way to deal with human waste and although it is certainly a fringe concept (even to me) I couldn't wait to actually try making one. It was four years from when I first learned of solar toilets until I found myself in a situation where I could, but that was just more thinking time for the idea to develop. Feedback from many conversations about this idea with various interested and uninterested individuals solidified my desire to create and try one out.

I built the test model from 25mm MDF (medium density fibreboard) that was epoxied watertight on the inside. This was then sheathed in 2-inch (50mm) expanded cell, polystyrene insulation board and then finally covered over in 3/8-inch (10mm) plywood for durability. I used a double-glazing of common Plexiglas for the window and spaced, as well as sealed, the two pieces of Plexiglas with a length of silicone tubing.

I will advise that actual use of such a device should not be attempted without first testing its capabilities; the very thought of what would be left if a device of this nature doesn't complete it's intended work encouraged me to look for alternatives. Less-than-savory clean-ups were not on my testing agenda and eventually I chose canned cat food as a worthy alternative to the 'real thing'. Confident that there would be performance testing issues if the device did not properly incinerate the deposits, I vastly preferred the idea of cat food clean-ups.

The first testing was less than dramatic and I produced nicely cooked, cat food burgers. In that the toilet hadn't achieved the solar incinerating temperatures that I believed were needed, I thought well of my foresight to use cat food. Venting modifications followed, along with other adjustments and replacements relating to the internal componentry of the box and it's glazing. I found that replacing a flat metal plate on which the cat food was cooking with an expanded metal screen made a big improvement in dry-out time and soon I had the feline treats thoroughly roasting to dry, little chips in a matter of hours. The end product was not to be mistaken for cat food or much else for that matter. It was dry, hard and odourless. There was no reason to be squeamish about picking it up with my hands. I believed that the time for testing the real thing had arrived.

Well, no need for detail here. Sabrina tells this story best, with all the emphasis of a sceptical, yet interested observer, that was intrigued by looking in on toasted cat food and unaware that an actual test was in progress the next time she checked in on things. The final results of the testing period were acceptable and the waste matter was, as with the cat food, dried out into toasty little chips. Also like the final tests on the cat food, the output from the 'human matter' was odourless and not offensive in any way (unless maybe you happened to know what it was).

Before I found a lab that would test the chips for bacterial life – obviously wanting to find none – I was informed that there was no chance of "that thing" ever being used as our primary toilet. I had reached a point of success that satisfied my curiosity but would have needed to spend more time to develop a way of speeding the dehydration process in order to use the unit for practical, everyday household needs. The cards were stacked against me so I let the solar toilet go to my list of 'one of these days, I'll be back to this' projects.

The back of the solar toilet allowed access to a clean-out drawer that would hold about a months worth of the dried wastes.

Application of the Knowledge

Solar Tracking

(As stated in Chapter Two, while not strictly a solar thermal design, this unit is worth mentioning as it can be adapted to devices as wide ranging as Photovoltaic (PV) panels to solar cookers.)

 Sabrina and I run our house on solar and wind power, with no connection to mains electricity. It is a choice we made, as the power line runs just a 100 yards from our house and through a long learning curve during which we received advice on alternative power living from anyone that cared to share a tidbit with us, we came to appreciate just how easy power is to consume and how not-so-easy it is to produce.

 Mechanisms that allow PV panels to follow the Sun from east to west each day can more than double the amount of power that the panels are able to produce. They come in a range of shapes and sizes and a range of prices too; the ones that we checked out seemed to always have a price range somewhere between expensive and really expensive. Then one day, with a conversation at a friend's birthday party with one of my U.S. Antarctic Program acquaintances, our power generating life changed for the better.

 I was told that the Program had given up on solar trackers at their installations because either the freezing conditions would lock them or the wind would destroy the tracker and subsequently the panels mounted on it, when everything came apart. Somewhere along the way however, somebody came up with the simple idea of a manually operated tracker that was repositioned as many times in a day as was convenient. Since field staff are never too far from their installations, at least two or three repositionings in a day were just about guaranteed. Sure, the amount of additional power wasn't quite as great as what was realised with an automated tracker, but these manual ones didn't fall apart when the winds got nasty rendering the whole installation useless either.

 My mental gears were turning after this conversation ... how could I put this piece of information to work at home?

My adaptation of the concept, shown in these photos carrying our PV panels, was built in an afternoon from scrap timber that I had lying around. We have more than doubled our PV's generating capacity with its use and since I work from home, I can move them along every couple of hours in the summer and more often in the winter, when additional accuracy counts as there is so much less sunlight available to us at that time of the year.

The supporting frame assembly is pinned to an open-frame deck with a 1/2–inch (12mm) lag screw. The assembly pivots on a large washer to give it a bit of 'float' for easy turning and a giant C-clamp holds the panels in place as they are repositioned. If we are away for the day we just rotate them to a neutral spot, centered on the midday Sun position in the sky and clamp them there.

Two nails driven into the center of the assembly assure tracking accuracy (see lower right photo, next page). By aligning the shadow from the PV panel's mounting bracket on the left set of panels to the nail nearest it, the Sun has approximately a 10–degree angle of incidence to the panels. A few hours later, once the Sun has passed by and the shadow from the mount on the right set of panels aligns with the nail nearest it, I rotate the mount so the shadow is again aligned to the left mount/nail.

(Remember, we're in the Southern Hemishpere. In Northern Hemisphere applications the unit would be moving from would be moving right to left.)

Manual tracker-mounted solar panels in their morning position ...

These nails, along with the shadow lines, provide a gauge for aligning the panels to the Sun with the least angle of incidence.

When the shadow from the right panel mount hits the long nail, it's time to move them along … get it?

… and later, after a day of following the Sun.

The adjusters that are used to set the panel's vertical alignment are evident in this view. Details of them appear on the next page.

Application of the Knowledge

Each pair of PV panels is mounted on five brackets, three at the base of the panels that allow a hinge-like motion (see photo lower right, previous page) and two that attach to retractable arms, connected part way up their sides, so the panels can be repositioned to follow the Sun's path (photo below). Whether the Sun is high overhead in the summer or low in the sky during the middle of winter, we can make adjustments to the panel's vertical orientation, enabling us to keep angle of incidence losses in that plane to a minimum.

I made the arms from lengths of square, hollow section aluminum by cutting the pieces to length, inserting the smaller section into the larger section and drilling holes through the pair. Adjusting the angle of each PV panel is as simple as aligning the holes in the inner and outer sections, inserting a bolt and securing it with a wingnut. The adjustment is made about once a month, which keeps the PV's close enough to a perpendicular alignment with the Sun to make a real difference in generating ability when compared to having them mounted in a fixed position.

Adjustable arms for mounting the PV panels allows for angle changes in the vertical plane to assist in maintaining a near-perpendicular alignment with the Sun. This simple item provides a great improvement in PV panel, power generating ability.

Passive Solar Construction

Encouraging a home and the site on which it is built to enter into a relationship with one another can be a challenging proposition. Passive Solar construction however requires the home and site to be viewed as one being, each representing an aspect of the relationship that can compliment and enhance the other. With collection, storage and transport of heat occurring by natural processes, there are however some trade-offs to be considered. When living in a rural or countryside setting, are people happy to make a sacrifice by siting their homes away from what may be the location's best view, in exchange for the best direction to collect free solar radiation? Will the homeowner be happy about eliminating large picture windows in order to avoid excessive heat loss in the winter? Other compromises appear as the design process progresses and the challenge to be faced by humans living within this house/site relationship is in regard to their tolerance of Nature having a say in the way they live their lives.

Factors to consider sometimes relate to the method(s) that will be used to collect heat in one area of the house, while shielding or protecting another area from overheating. Other factors involve the physical characteristics of the site and as highlighted above, choices that can lead to the best passive solar results obviously have an impact on the home's design and the materials that are used in its construction. When all factors are taken into account, the final design may not be exactly in line with what the owners originally thought their home would look like. The smart money is on those that are willing to make the compromises; humans don't *have* to control *everything* in their lives.

When thinking about the different ways in which passive solar principles can be put to use in a home, there are basically five systems to consider:

Solar windows. Windows that allow sunlight to enter your house directly through its windows and turn into heat are sometimes referred to as solar windows. Some of the heat from such windows is used immediately, while floors, walls, ceilings, and furniture store excess heat and release it later when the sunlight is gone.

Solar chimneys. By taking advantage of the natural process of convection, well-designed chimneys can be used in the summer to draw cool air into the home while exhausting hotter air from the top of the chimney, creating a natural type of air conditioning. If the chimney is designed with ducting that leads back into the house, warmed air collected within can be directed into the house in the wintertime by fans – not passive, but effective.

Mass walls. When the mass for absorbing the Sun's heat is located inside the home, it is called a mass wall. The most effective mass walls are painted a dark color to assist them in heating up as sunlight passes through the windows and strikes them. Heat from a mass wall is conducted through the wall and radiates into the house. Natural convection processes also take effect, spreading the warm within the room after the room's air is warmed by contact with the wall's inner surface.

Solar roofs. Like solar walls, but collecting heat on the outside of the home. Most solar roofs use water in large black plastic bags or open ponds to absorb heat from sunlight during the day. The water stores the solar heat, which in turn is conducted through the ceiling and radiated to the house below. Insulating panels can be used to cover the ponds at night to reduce heat loss.

Sun rooms. Rooms such as attached greenhouses, sun porches and solariums are possibly the oldest form of a household, passive solar heating system. They give the house an extra, solar-heated living space and act as a buffer zone between the indoor living spaces and the outdoor weather extremes.

As mentioned in the opening paragraph, designers need to be aware of the surrounding landscape in order to take advantage of the best times of the day for solar exposure. A house is an *extremely* fixed, solar collector. If the site sometimes has too much solar exposure, say for example complete afternoon sunlight that can cause overheating, then the designer can factor in either landscape features like large trees to shield the home from the Sun or to build in shielding mechanisms as part of the house itself.

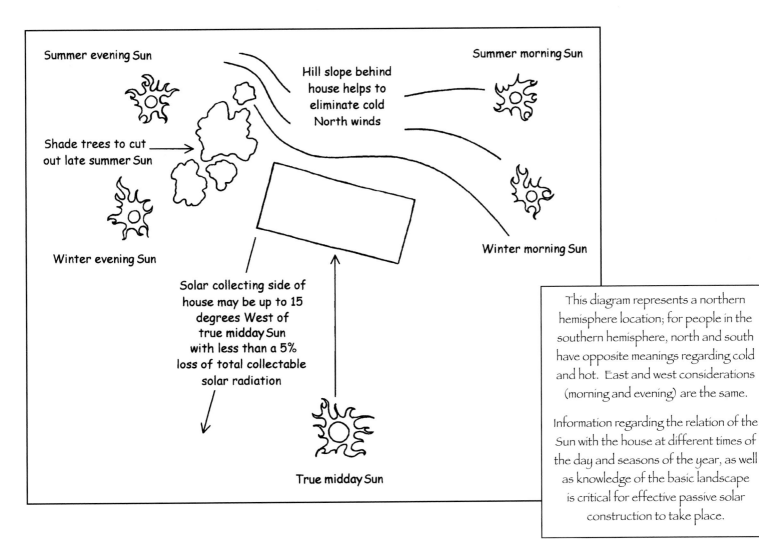

Summer evening Sun

Hill slope behind house helps to eliminate cold North winds

Summer morning Sun

Shade trees to cut out late summer Sun

Winter evening Sun

Winter morning Sun

Solar collecting side of house may be up to 15 degrees West of true midday Sun with less than a 5% loss of total collectable solar radiation

True midday Sun

This diagram represents a northern hemisphere location; for people in the southern hemisphere, north and south have opposite meanings regarding cold and hot. East and west considerations (morning and evening) are the same.

Information regarding the relation of the Sun with the house at different times of the day and seasons of the year, as well as knowledge of the basic landscape is critical for effective passive solar construction to take place.

Application of the Knowledge

Designing Solar Gain into a House

Think of the rooms in the house you live in now. Is there a room that gets very warm during the daytime because of a large window or set of glass doors? Is that same room also the one that gets very cold on a winter's night? This room is feeling the effects of uncontrolled, direct solar gain and radiated heat loss. Thermal storage such as the previously mentioned mass walls and other mass bodies can help to alleviate these heating and chilling cycles by inducing a **thermal flywheel** effect.

A thermal flywheel amounts to nothing more than bodies of dense materials that are capable of soaking up and holding heat inducing a delayed effect in regard to spaces heating up and cooling down, in a manner similar to the way a mechanical flywheel smoothes out the fluctuating input from a motor. Some common mass bodies used in home construction are:

- Concrete floor slabs - these can even be thickened to provide larger storage capacity where the sunlight is able to make direct contact with the floor;

- Masonry walls - solid internal walls or *externally insulated* exterior walls that capture heat from the room's air;

- Water Chambers - Containment vessels of all sorts can be used to hold water, which as we know, is one of the best naturally occurring heat containment materials.

A building in which the solar radiation first heats an isolated object that then transfers the heat to the interior living spaces is considered to be using an indirect gain system. The use of thermal flywheels consisting of mass walls in either masonry construction or water containment vessels that are placed in a position between exterior glass panels and the living space of the room is typical in an indirect gain system. The mass wall is heated by direct gain and internal conduction. Heat transfer to the room then occurs either by radiation from the heated mass to materials in the room or through vents placed in such a position that a thermosiphon, as discussed way back in chapter one is induced.

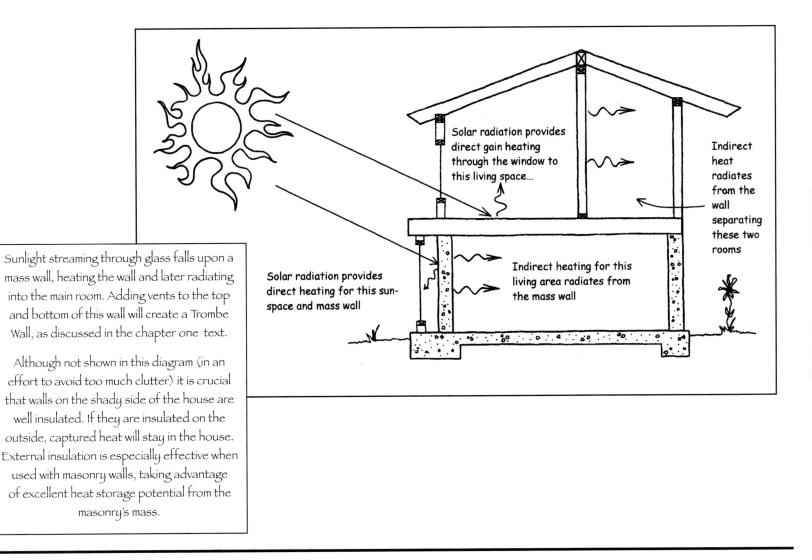

Sunlight streaming through glass falls upon a mass wall, heating the wall and later radiating into the main room. Adding vents to the top and bottom of this wall will create a Trombe Wall, as discussed in the chapter one text.

Although not shown in this diagram (in an effort to avoid too much clutter) it is crucial that walls on the shady side of the house are well insulated. If they are insulated on the outside, captured heat will stay in the house. External insulation is especially effective when used with masonry walls, taking advantage of excellent heat storage potential from the masonry's mass.

Solar radiation provides direct gain heating through the window to this living space...

Indirect heat radiates from the wall separating these two rooms

Solar radiation provides direct heating for this sun-space and mass wall

Indirect heating for this living area radiates from the mass wall

My Own Use of Passive Solar Construction Concepts

Stone House

The house built by my partner Sabrina and I was designed around the basic principles of passive solar design. It is sited to maximize our ability to collect warmth from the Sun in the winter and has massive stone walls to retain the collected heat. The roof eaves are large enough to shade these walls and keep them cool in the summer and on its permanently shaded side, the house is partially buried to assist in moderating the internal temperature. It is also very well insulated both above and below ground levels.

Neither of us had ever lived in such a structure before building this one, so it was something of a leap of faith in all that we had read about passive solar construction to push on with it. Truth be told, the house is remarkably comfortable year round. Requiring heat from a potbelly stove on only the coldest winter days and remaining comfortably cool in the summer, I could hardly image a better place to live.

For those that might be more interested in self-building or the slipform method of construction, I highly recommend purchasing a copy of my book *Stone House – A Guide to Self-Building with Slipforms*. Well, of course I would recommend one of my own books but it really does have more information on slipforming than can be found in any other book about the subject, as well as many related topics that are of interest to self-builders.

Siting Considerations

The two main factors that Sabrina and I took into account when first considering our house's siting, were the house's position in relation to the Sun and protection from some local weather considerations. We live in a valley setting, surrounded by hills with the mouth of the valley opening on to a large freshwater lake that is separated from the Pacific Ocean, just five miles away, by a thin spit of sand dunes and marginal farmland. Living in a hilly district can pose its share of challenges for passive solar designers. Sunrise and sunset do not correspond with the rising and setting times for the general region and weather patterns (particularly those relating to wind direction) are influenced by the hills. Winds, which influence a large part of New Zealand's climate, can do odd things when exposed to gullies, ridges, and hills after coming across thousands of miles of open ocean.

The weather conditions took first precedence in our minds. Our region of New Zealand has a prevailing wind from the southwest that varies from breezy to positively *fierce*. I've seen seriously harsh wind conditions when I was in the Antarctic, and these southerlies can be every bit as bad as what Antarctica can to dish up, just not quite as cold.

The conditions of the Southerly winds are often worsened by the addition of a cold rain that can accompany them. Coming into the valley with winds ranging in speed from 5-20 metres/second (I don't see a point in converting these wind speeds to imperial, just understand that I'm talking about horizontal rain), the rain is potentially as damaging to a home as the wind. Lucky for us, we found a nice spot tucked around from the property's southwest corner, on a ridge-backed hill that is oriented east-west. It provides an excellent barrier to these ferocious winds, and just happens to face due north. Perfect.

With the site chosen, our concerns turned toward the solar orientation of the house. Because we are in a valley, our site's "first Sun" is typically 3 or 4 hours after the actual sunrise for the region. Later in the day, we lose the Sun behind the ridge on the opposite side of the valley, several hours before the region's actual sunset. We decided to base our house's solar orientation on the Sun's mid-sky position, in relation to what we experience as a 'Sun day', on June 21st, Midwinter's Day here in the southern Hemisphere (the corresponding day in the Northern Hemisphere is December 21st).

To learn just where in the sky this position was achieved, we had to arrive on the property early enough in the morning on June 21st to beat the Sun coming over our property's northern ridge. Then we needed to stay late enough to see it set on the west arm of the valley, to discover that position. The mid-sky position would be right between these two points.

The sunrise part was no problem. The house site's sunrise happened at 10:05am. There's no need to say, not many early birds are still out catching worms at that time of the day. Sunset was also no problem – 4:00pm. We do not exactly have what most people would consider to be a long day, so we obviously need to make the most of the sunlight that is available. We picked our mid-sky position based on sound, logical judgment of the facts that we now had, and then tweaked it just a bit to give us a nicer outlook from the main living room windows. We finished the exercise with a siting that is about 15 degrees west of true north; not ideal in the textbooks of passive solar design, but perfect based on our local conditions.

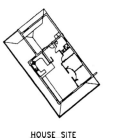

HOUSE SITE

East

West

An excerpt from our Site Plan showing the house oriented approximately 15 degrees west of true north.

Floor Slabs

Underneath a typical concrete floor slab is a layer of plastic moisture barrier, then some gravel infill and then the solid bearing earth. In the early days of slab-type floors, the gravel infill layer was considered to offer an adequate insulating effect. Air is trapped in this layer of material, and trapped air is an insulator, therefore there is a beneficial effect; the problem with relying on this air-filled gravel layer as floor slab insulation is that it is very inefficient. The slab loses any gained heat very quickly and many people with older homes complain that the bare concrete floor is cold on their feet. This happens because the slab and the ground below it are seeking a state of thermal equilibrium. Of course the ground, fuelled by the thermal mass of the entire planet, wins out in such a competition when pitted against a measly little floor slab (relatively speaking). The floor slab loses any heat that it may have and adopts the temperature of the ground below it.

But it doesn't have to be that way; a concrete floor slab can work to enhance the comfort of any home that is being designed to have one. By using under-slab insulation, floor slabs can be made to enhance the passive solar effect. Once insulated, the slab is thermally isolated from the ground temperature below. With an insulated slab such as this, the floor temperature can be controlled (mostly) by the occupants; the temperature of an insulated slab will generally be that of the house's room temperature.

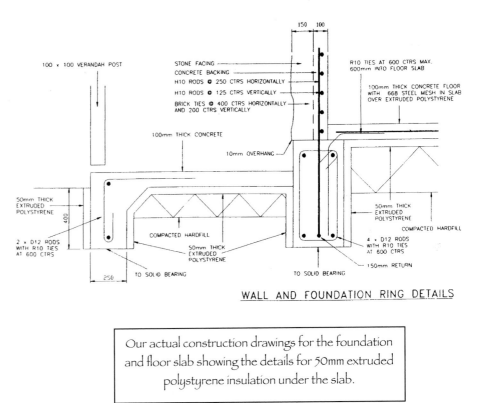

WALL AND FOUNDATION RING DETAILS

Our actual construction drawings for the foundation and floor slab showing the details for 50mm extruded polystyrene insulation under the slab.

Application of the Knowledge

Working with the annual Sun tracking patterns in the sky, window and door openings that have been situated properly allow portions of the floor slab or exposed walls to receive direct solar radiation at the right times of the year. This benefits the heating of the house and, when properly managed year round by also shading the slab in the summer, a thermal flywheel will be created. Mass walls and insulated floor slabs becomes an integral part of the homes climate control systems if they are designed properly and insulated to hold their desired temperature.

Polystyrene

As mentioned back in the facts about insulation, polystyrene insulation comes in several types, named after the different methods in which the liquid polystyrene is formed into sheets. Extruded polystyrene is without a doubt the best choice for all subterranean insulation requirements. Be sure to only consider the *extruded* variety when working with underground applications; *expanded* cell or 'beadboard' polystyrene, the type that crumbles into millions of little balls when scrubbed on the ground, will not suit for underslab insulation. It has the ability to absorb moisture, which renders it useless for insulation and also leaves it open for rot and physical degradation. No one wants this happening underneath their house.

On the other hand, the cell structure of extruded polystyrene will not allow water infiltration and it can bear extremely heavy loads; I have read of this stuff being used underneath motorways. It is certainly not the cheapest option but as with so many good things, the payoff is in the long term. Between great insulating ability, fantastic structural strength and an additional element of waterproofing, it's hard to go wrong. I speak from the heart when I suggest: spend the extra money, use extruded.

How much insulation is required to make this concept work? We used a 1¼-inch (30mm) thick product under our entire slab and have achieved great results. Extruded polystyrene is available in up to a 4-inch (100mm) thick material but the price starts getting really thick as well. For us it was a balance of performance vs. budget. If there had been a little more money available in our budget I would have used 2–inch (50mm) for the under slab insulation but, as I said, we have a slab that performs very well with 1¼-inch (30mm) insulation under it.

Our entire floor slab, including the verandah, has the same thickness of extruded polystyrene to protect it from the heat sink effect experienced by concrete slabs in direct contact with the ground.

Below Grade Walls

Any below grade walls can take advantage of this insulating method as well. By insulating them on the outside, thermal mass can be added to the interior of the home. An advantage to subterranean walls is that they are exposed on the outside to only ground temperature 24 hours a day. While the air temperature swings up and down from day to day, and sometimes drastically within a day, the ground temperature only a foot or so below the surface changes very little. This means that regardless of whether it's a blistering hot summer day or a freezing cold winter's night, below grade wall insulation is only feeling the constant ground temperature of about 55°F (15°C). No wonder my grandmother's basement felt so nice on a hot summer afternoon, even if it wasn't insulated.

Because of the way that we nestled our house into the hill, the two rear walls were to be earth bermed up to 3 ½ feet (1400mm) above the level of floor slab. From that point up, the walls were above ground and made from stone. The portions of the walls that were buried were cast from concrete, using the slipforms for the casting moulds.

There was no messing around back here with site mixed concrete either; these sections were going to consume nearly 20 cubic metres before we got up to where the stone was going. We called in the Ready Mix concrete trucks and their concrete pump truck to do this job (see photo right).

As with the floor slab, the subterranean portions of our exterior walls used extruded polystyrene to insulate them from ground temperature, but we had to treat this situation a bit differently. Unlike insulation under a floor slab, buried walls are potentially exposed to both moisture and burrowing animals, so we had to ensure that the insulation was well protected and the whole arrangement made waterproof.

There is a wide variety of water sealing products for subterranean wall applications on the market. They range from sticky, gooey stuff called "Black Jack" that is painted over the walls, big impermeable membranes that are glued in place and chemicals that are mixed into a concrete plaster while the mix is being prepared. What is really wanted for underground walls is a waterproof situation that never allows water to get as far as the waterproofing. This can only be done well with layers of different materials.

We liked the idea of waterproof, concrete plaster as a starting point, the first layer if you will. Our main attraction was that we thought it was an easy method of application and that it was rugged enough to be a strong protection from burrowers like rabbits, as well as a 'last resort' protection from groundwater.

A back wall, with the forms for its last lift still in place. The nuggety texture of the poured concrete on these walls was a perfect key for the plaster coats that would follow.

Application of the Knowledge

The first plaster coat, applied directly to the rough-cast walls, served to also secure a strip of 13mm wire mesh that would be folded over the top of the polystyrene after it was applied. The polystyrene, that represented our second layer of waterproofing, was pre-moulded with a diamond shaped scoring on its surface, so an outer layer of plaster would adhere to it easily. The embedded mesh eventually formed a well-secured lath for the top of the final plaster coat to reach up to the stones.

For securing the sheets of polystyrene to the plastered concrete walls, we used a glue product formulated especially for use with polystyrenes. Most regular construction adhesives will 'eat' polystyrene so finding the right product is critical. We bought ours from the same place that supplied us with the insulation.

Insulation is held in place by 2x4's while the glue cures. The wall has already received its first waterproof plastering coat.

13mm mesh, embedded in the first plaster coat and to be used as a lath over the top of the polystyrene for the final plaster coat, is clearly visible.

When the glue holding the polystyrene to the walls was cured, we applied the final waterproofed plaster layer. This layer covered the entire polystyrene surface, starting from down at the foundation and drainpipe and continuing up to just below where the stonework begins.

At the base of the walls a 6-inch (150mm) polythene drainpipe was laid into a bed of gravel. This drain directs any water that manages to get down that far, out through holes in the retaining walls. From there, the water will run into the storm-water drainage system (photo, right).

Finally, once the second plaster coat was cured the whole arrangement was backfilled with large round gravel to nearly surface level (lower photo, left). The gravel was then covered over with a polypropylene geotextile fabric to keep surface soils from settling down into it. To finish off, the surface soils were placed about a foot deep, then seeded and planted-in with decorative grasses.

Application of the Knowledge

Roof Overhangs

I don't know what's happening in the rest of the world, but here in New Zealand, roof overhangs seem to be disappearing. It may be a design trend or, because with a smaller the roof there is less money to be spent on roofing materials, it may be an economic trend but it definitely is happening. At the same time, more effort is being put in by manufacturers to design window systems that have a greater ability to keep water from leaking in past their weatherproof flashing strips, water infiltration in buildings has been the topic of several consumer focused television shows, and I've been barraged lately with advertisements regarding all the latest glass technology; films, coatings and in-manufacture treatments that are designed to filter out harsh ultraviolet light in order to stop furniture and carpets from fading (or being destroyed) and also to keep the interior spaces of home nice and cool in the summer.

Could the four problems highlighted here, leaking windows, leaking buildings, ultraviolet damage and overheating in the summer be resolved if ample roof overhangs once more became "the trend"? Although it must be remembered that the purpose of the roof is to protect the house and allow rain and/or snow to be shed from it, if properly designed the roof can also be our most considerable ally in the battle against internal ultraviolet light damage and overheating.

Provided the local summer/winter solar angles are known and local weather conditions are considered, the minimum roof overhangs can be established to provide a passive solar designer with some control over these factors. Taking into account the overall design characteristics of the house, the overhangs may be reduced below this point, but whenever they are, in most situations the roof loses some of it ability to do its job. Without the roof working for us 'full time' then it can only be expected that we are going to experience difficulties somewhere else in the building, like leaking windows or fading carpets.

For Sabrina and I, the minimums were determined with two criteria in mind. First, we wanted 100% solar exclusion on the north side of the house at midday, December 21st. That is Mid-summer's Day, the longest day here in the Southern Hemisphere (people in the Northern Hemisphere can look for the same 'maximum Sun' day on June 21st). Because of the nature of the Sun's path in the sky, rising to and then descending from its mid-Summer position, 100% exclusion on this day will effect a condition where there is the least potential for ultraviolet or overheating problems for a period beginning one month before and carrying on until one month after Mid-summer's day. The overhang for the North-facing side was calculated to 56 inches (1400mm).

Our second criteria relating to our roof overhangs was to provide as much protection as possible from the driving Southerly rains that I've mentioned. These rains hit us on the Southwest side of the house after wrapping around the West end of the hill that the house is built into. There is no possibility for 100% exclusion because these rains generally come in sideways, so our parameter for the overhang on this side was to go to the maximum allowable extension, based on engineering guidelines. That result was a full 6 feet, 7 inches (2 metres).

As can be seen from the shadow lines on these two photos, we have 100% solar exclusion at midday, on December 21st (left photo) and there is very little potential for internal u.v. damage or overheating. The right photo shows the same basic view as seen in the left photo but is taken at mid-winter. At this time of the year we go for maximum solar gain from sunlight in direct contact with the darkened concrete floor inside.

Application of the Knowledge

Finale

From this point, the process of designing, making and using solar thermal devices goes to you, the reader; the designer. Will a state of equilibrium be achieved between solar designers, the production and use of their devices and the functioning everyday world of practical people that are all around us?

Perhaps an answer lies in a rhetorical application of the theories that we have already discussed. In a way reminiscent of captured heat exceeding an insulating material's U-value, if a core group of individuals generates enough enthusiasm and applied use of such principles and theories as we have discussed, the knowledge of using solar thermal devices will begin to seep out from this 'heated' core group, penetrating the insulated knowledge base of the population at large. In the way that heat can overcome an insulating material's grasp, my greatest wish in assembling this collection of information is that the knowledge contained here will awaken a creative genius in readers that they cannot contain. The knowledge will be spread in such a way as only those with a thorough understanding of what is going on beneath the surface of a subject can do. If this book can spark the creation – even invention – of new ways to harness the power of our nearest star and a greater acceptance of solar thermal devices throughout the world, my wish will be granted.

I've said it before, I'll say it again … knowledge is power. We have the power of the Sun and the knowledge of principles that contribute to good solar designing, let's not turn away from the opportunity to make a difference to our world.

My best wishes go out to you and your own solar thermal designs,

A Chronology of Passive Solar Achievement

B.C.

5,000,000,000 B.C. – the Sun begins to form from a cloud of gasses in the primordial Milky Way Galaxy.

Socrates (469 – 399 B.C.) defines principles for using passive solar design in houses.

Aristotle (384 – 322 B.C.) suggests a method of desalinating seawater through solar heated evaporation.

Archimedes (287 – 212 B.C.) creates solar focusing war mirrors at Syracuse.

1st Century A.D.

10	Large baths in Rome utilise dark-colored pottery to warm the water.
20	Chinese document use of burning mirrors to light torches for religious ceremonies.
37	Greenhouses are used to grow cucumbers for Tiberius Caesar.
100	Heron of Alexandria constructs a solar device for water concentration.
100	Italian historian, Pliny the Younger builds a passive solar summer home in Northern Italy using thin sheets of mica for windows in one room.

500 A.D.

528	Roman Law is collected under a single constitution known as the Justinian Code, which contains articles to protect light access to rooms on the sunny side of houses and public buildings.

1400's

1476   The ancient Incan city of Machu Picchu is established in the Peruvian Andes Mountains, with massive east-facing stone walls believed to be intentionally designed to gain heat from the morning Sun and to retain it throughout the day.

1500's

1515   Leonardo da Vinci conceived a design for a parabolic mirror to concentrate heat for industrial applications.

1560   A solar still is constructed for the French surgeon Amroise Pare.

1561   Alchemists submerge flowers in a water-filled vase, which was placed at the focal point of a spherical mirror; the concentrated solar heat causes the essence of the flowers to diffuse into the water.

1600's

1610   Mobile burning mirrors become common for use by scientists.

1615   Solomon de Caux builds a solar water pump made of glass lenses, supporting frame and an airtight metal vessel containing water and air that powers a small fountain.

1646   Athanasius Kircher builds a solar furnace that melts lead and distills water in a jar. He also designs plans for a mirrored 'heliostat weapon'.

1697   Grao-Duque Cosme III, of Toscana, uses sunlight and convergent lenses to burn a diamond.

Sir Isaac Newton uses a system of concave mirrors to focus solar rays and develops the Newtonian reflector type of telescope, which uses a parabolic mirror to bring starlight to a focus with the aid of additional optical components.

1700's

Holland develops greenhouses with South facing sloped glass walls.

European aristocracy use walls to store heat for ripening fruit.

1747 Frenchman George-Louis Leclerc, concentrates sunlight using 140 plain mirrors at a distance of 213 feet (65 metres) to ignite wood and later at 328 feet (100 metres) to melt lead.

1767 Swiss scientist Horace de Saussure creates a series of glass hot boxes to investigate solar warming of enclosed spaces.

1770 French scientist Marc Ducarla Bonifas adds insulation and mirrors to reflect more light into Saussure's design.

1774 Using two focusing lenses, Antoine LaVoisier builds a solar furnace that is recorded to have reached temperatures in excess of 3,000°F (1750°C).

1800's

1826 John Ericsson builds a hot air engine powered by the Sun.

1837 Astronomer Sir John Frederick Herschel uses an experimental solar box to cook food on a South African exploration at Cape of Good Hope.

1861 Auguste Mouchot patents his first solar engine.

1869 In his book, *Moyens de Communications avec les Planètes,* Frenchman Charles Cros speculates on using burning mirrors to communicate with Martians by burning messages into the surface of Mars.

1869 Mouchot publishes his experiments with a 1.5 Kw engine built in Algeria.

1870 John Ericsson develops what he calls the first solar powered steam engine, dismissing Mouchot's as "a mere toy."

1874 J. Harding and Charles Wilson create a solar desalination plant for drinking water at a nitrate mine in the Atacama Desert at Las Salinas, Chile. Producing over 5,000 gallons (24,000 litres) a day, it operated for over forty years and outlasted the mine.

1878   Mouchot presents a steam engine coupled to a refrigeration device at the Paris Exhibition that used the solar heated steam, rerouted through a condenser, to cool the air inside an insulated box.

1878   Abel Pifre, Mouchot's assistant, demonstrates his solar powered steam engine, powering a printing press at the Paris Exhibition.

1878   English diplomat William Adams designs and patents a solar engine using a system of heliostats (mirrors) and a central boiler that will later become known as the Power Tower concept, at his compound in India.

1880   E.J. Molera and J.C. Derbrain obtain solar boiler patents in Germany.

1881   Edward Morse applies for solar space heater patent. The device is very similar to what is later known as a Trombe Wall.

1882   W. Calver of the United States patents the first of two solar engine designs.

1884   Dr. Samuel P. Langley experiments with hot boxes on Mount Whitney, California.

1885   Charles Tellier obtains a solar steam boiler patent using ammonia as the liquid and presents the first example of a non-concentrating, non-reflecting solar powered motor.

1886   John Ericsson produces a solar engine that uses a parabolic trough.

1889   Tellier increases efficiency of the collectors by enclosing the top with glass and insulating the bottom and publishes the results in *The Elevation of Water and the Solar Atmosphere*.

1891   Clarence M. Kemp of Baltimore designs and markets the first commercial solar water heater, "The Climax."

1896   C.G.O. Barr receives patent for a very large solar engine using a semi parabolic mirror array mounted on railroad cars on a circular track with a fixed boiler at the focus of the system.

1897   Kemp's "Climax Solar Water Heater" and others are responsible for heating water in 30% of homes in Pasadena,California.

1899   English engineer Aubrey Eneas patents his solar engine.

1900 – 1909

1901 Aubrey Eneas promotes his solar powered water pump at Edwin Cawston's Ostrich Farm in Pasadena, California.

1902 F. Walker patents a combined solar and artificial hot water heater, a very early, hybrid water heating system.

1905 H.E. Willsie and John Boyle build a solar engine producing five Kw of power and patent their "Solar Power Plant".

1905 E.P. Brown and Carl Gunther obtain patents for solar steam boilers.

1907 W. Maier and A. Remshardt obtain solar steam boiler patents.

1909 William J. Bailey designs an insulated solar hot water heater for Carnegie Steel, creating The Day and Night Solar Water Heater Company.

1910's

1912 Frank Shuman's solar powered irrigation pumps successfully deliver commercial quantities of water at Meadi, Egypt.

1916 Charles G. Abbot (1872–1974), builds a solar oven, which used circulated oil as its heating medium and achieved 350°F (176°C).

1917 Shuman and Boys obtain patent for their solar engine.

1920's

1921 A solar engine is patented by W.J. Harvey.

1924 Famous rocketeer Robert Goddard receives the first of his five patents in the field of solar heated devices.

1928 A solar engine is patented by L.H. Shipman.

1930's

1930   H. Delecourt develops and constructs a solar engine using ethyl chloride.

1930   D.H. Drane develops and constructs a solar engine using ammonia.

1931   G.W. Dooley receives patent for solar engine.

1932   W.L.R. Emmet patents a mirrored solar engine inside a vacuum envelope.

1933   George and William Keck build America's first modern passive solar home at the Century of Progress Fair in Chicago, called the "Crystal House".

1935   F.A. Gill receives patent for solar engine.

1936   Charles G. Abbot exhibits a 1/2 horsepower solar operated steam engine at the International Power Conference in Washington D.C.

1938   Arthur Brown designs a passive solar schoolhouse in Tucson, Arizona.

1940's

1941   The first report of an Australian domestic solar water heater built and installed at Meringa Station near Cairns, North Queensland, Australia appeared in *The Cane Growers Quarterly Bulletin* of July. The system had 20 square feet (1.9 sq metres) of collector area and held 40 gallons (180 litres) of water.

1944   Maria Telkes designs a portable, solar sea water distiller for pilots shot down over the ocean capable of producing one quart of water per day.

1947   Libbey-Owens-Ford Glass Company publishes *Your Solar House*, profiling forty-nine designs by prominent U.S. architects.

1949   An experimental solar oven designed by Felix Trombe is built at Odeillo in the French Pyrenees.

1950's

1950   MIT holds symposium on solar house heating in Cambridge.

1952   Iran formally breaks relations with Great Britain in a dispute over oil.

1953   University of Wisconsin holds a symposium on solar energy utilization.

1953   Frank Bridgers designs Bridgers-Paxton Building, listed in National Historic Register as the world's first commercial solar heated office building.

1953   Australian firm S. W. Hart and Co. Ltd. started making solar water heaters. The company now trades as Solarhart and is sold in 72 countries.

1954   UNESCO and the Indian Government sponsor a Symposium on Solar and Wind Power in New Delhi.

1954   The Association for Applied Solar Energy is founded in the U.S.

1955   First World Symposium on Solar Energy is held in Phoenix, Arizona.

1957   George Lof designs the "Umbroiler", a parasol-like foldable parabolic concentrator for cooking.

1958   Felix Trombe invents the Trombe wall.

1959   D.S. Halacy Jr. publishes *Solar Science Projects*, featuring a solar oven.

1960's

1960   Harry Tabor of the National Physical Laboratory in Israel develops a solar powered turbine, which uses the heavy gas monochlorobenzene as an operating fluid.

1961   The Social and Economic division of the United Nations organizes its Conference on New Sources of Energy in Rome.

1962   Natick Labs solar furnace capable of achieving 7,000°F (3871°C) is built to simulate radiation burns and to experiment with pigs wearing sunscreen lotions that might protect soldiers.

1970's

1973   A 100 year commemoration of the solar distillery in Chile is the backdrop for the formation of the Latin American Association of Solar Energy.

1973   The University of Delaware builds "Solar One," a PV/thermal hybrid system In addition to providing electricity, the photovoltaic arrays are used as flat-plate thermal collectors; fans blow warm air from over the array to heat storage bins.

1973   First embargo of oil to the United States begins.

1974   The International Energy Agency is formed to facilitate the development and transfer of energy technologies that promote the energy security, environmental protection and economic development of its Organization for Economic Co-operation and Development (OECD) member countries.

1975   First Chinese National Solar Energy Congress is held.

1975   Italian Professor Franciain Sant'Illario heats a steam boiler using 120 mirrors generating 932°F (500°C).

1976   65Kw solar power station in Odeillo, France goes online as the first solar power to be connected to a main grid.

1977   The U.S. Energy Research and Development Administration, a predecessor of the U.S. Department of Energy, launches the Solar Energy Research Institute, a federal facility in Golden, Colorado, dedicated to finding and improving ways to harness and use energy from the Sun.

1978   New Mexico's Solar Rights Act of 1978 is passed, allowing property owners to create solar easements for the purpose of protecting and maintaining proper access to sunlight.

1978   The Victorian Solar Energy Research Committee installed a demonstration unit for solar water heating at the University of Melbourne's Beaurepaire Pool, beginning an 18 month performance monitoring period.

1979   A solar hot water heater is added onto the West Wing of the White House under President Carter. It consists of 34 collectors and supplies 75% of the White House's hot water needs.

1979   Second US oil embargo begins.

1979   The Solar Trade Association (Solar Energies Industries Association) is established in Washington, D.C.

1980's

1980   The first solar cell power plant was dedicated at Natural Bridges National Monument, Utah. The $3 million photovoltaic system had 266,029 solar cells mounted in 12 long rows producing a 100 Kw output.

1980   The U.S. Energy Security Act effectively shuts down the national solar program and orders the Carter solar panels to be removed from the White House.

1981   The former White House panels are installed at Unity College in Unity, Maine.

1982   Power Tower One, a solar-thermal power plant collaboration between the U.S. Department of Energy and a consortium of corporations starts operation near Barstow, California.

1983   Wisconsin "Right to Light" law enacted to preserve sunlight for urban gardens. Michigan and Arizona follow suit.

1987   Solar Cookers International is founded July 11 in a spare room at the University of the Pacific in Stockton, California by a group of educators at the urging of one Beverly Blum.

1987   Gruppe ULOG builds the first of sixty Solar Hybrid Kitchens throughout Kenya, Sudan, Cameroon, Peru and India.

1988   The Solar One plant is deactivated.

1990's

1992   China claims to have 100,000 solar cookers in use designed by the Henan Academy of Sciences.

1994   A hospital in a Tuen Mun, Hong Kong, installs 1033 sq feet (96 sq metres) of solar panels designed to serve a hydrotherapy pool.

1994   World Conference on Solar Cookers is held in Costa Rica.

1996   Southface Energy Institute opens a state-of-the-art demonstration building, the Southface Energy and Environmental Resource Center in Atlanta Georgia, to provide a permanent showcase for innovative technology and a resource for builders and homeowners throughout the Southeastern U.S.

1996   Solar energy is used to heat the water for swimming events at the Atlanta Olympic Games.

1996 Power Tower Two a 10 MW, second-generation solar thermal power station begins operation at the same site as Power Tower one in Barstow, California. Solar Power Tower Two is made of 1,926 heliostats focusing on a 300 foot central tower. It is capable of providing power for 10,000 homes.

1997 The U.S. Department of Energy announces its Million Solar Roofs Initiative, a program to facilitate the installation of solar energy systems on one million U.S. buildings by 2010.

1997 Evacuated Heat Pipe Solar Collectors installed for the in-flight catering building at Singapore's Changi Airport is the largest ever solar heating project in South East Asia.

1997 Through building code AS/NZS 4445.1:1997, Australia and New Zealand implement a performance rating procedure using indoor test methods for solar heating as applied to domestic water heating systems.

1998 Task 24 "Solar Procurement" is begun by the International Energy Agency in Stockholm, Sweden. Sweden, Canada, the Netherlands and Denmark agreed to a joint effort to create a sustainable, enlarged market for active solar water heating systems.

1998 Ralph Goodale, Minister of Natural Resources, Canada, inaugurated the first installation of a solar air heating system under Canada's Renewable Energy Deployment Initiative (REDI). The event took place at Tapis Coronet Inc. – Coronet Carpets Inc., the first company in Canada to benefit from the business incentive portion of the REDI program.

1998 A community swimming pool complex in Kwai Chung, Hong Kong, installs 3,552 sq feet (330 sq metres) of solar collectors to heat the pool's water.

1999 The EU Ekoviikki solar project in Finland is launched. Consisting of 8 building-integrated solar heating systems with a total area of 13,433 sq feet (1248 sq metres), the systems provide domestic hot water and in some cases space heating through in-floor systems.

1999 Power Tower Two is taken out of service.

1999 The first central solar heating plant that uses an aquifer thermal energy store is completed in Rostock, Germany. The unit stores thermal energy in wells up to 98 feet (30 metres) deep and services 108 apartments.

1999 The International Energy Agency (IEA) celebrates 25 years of participation and development of solar power.

Appendix 1

2000 and Beyond

2000   California's Governor Davis signs State Senate Bill 1345 that provides funding of up to $750.00 per home for solar water heating systems as well as distributed generation systems. The Solar and Distributed Generation Grant Program is to be administered by the California Energy Commission.

2000   The world's largest solar cooking system started functioning at Taleti, near Mt. Abu, India. Installed by the Brahmakumaris organization with assistance from the Ministry of Non-Conventional Energy Sources, the system is designed to prepare meals twice a day for 10,000 people and saves the community up to 105 gallons (400 litres) of diesel per day.

2001   One million square metres of solar collectors are installed throughout the European Union.

2002   Australia, Austria, Belgium, Canada, Denmark, European Commission, Finland, France, Germany, Italy, Mexico, Netherlands, New Zealand, Norway, Portugal, Spain, Sweden, Switzerland, United Kingdom and the United States are formally detailed as participants in the IEA's alternative power programs.

Beyond.

Where does 'beyond' take us? We are now living in the beyond; the timeline is continuing all around us and as can be gleaned from this historical record, by the end of the 20th century the use of passive solar devices had spread to all nations of our planet. Everywhere around the world people are using the Sun's warmth to improve not only their own lives by utilising this free, non-polluting resource, but the lives of the future generations that will inherit the legacy of what we leave behind. If our generation accepts solar and other non-polluting sources of energy, our descendants will inherit a clean, healthy place to live.

International Astronomical Union (IAU) Nomenclature

Although a wide range of literature, from common magazine or newspaper articles to university level textbooks, will present celestial bodies such as the Sun, the Moon and the Earth in an un-capitalised form, the capitalization of these objects is in keeping with the IAU guidelines for astronomical nomenclature. Sun, Moon, Earth, etc. are proper names for an object and therefore should be capitalized, as is accepted practice with all proper names such as Tomm, Fido, Milwaukee, etc. it is predominantly the source of followed habit that finds the names of these objects spelled in an uncapitalized format.

The IAU also advocates the elimination of the word "the" from preceding a celestial object's given name. This is in keeping with correct written and spoken grammar; I am not referred to as "the Tomm" and we do not refer to other celestial objects as "the Mars" or "the Venus" therefore we should not refer to the Sun, the Moon or the Earth in this fashion either. Again, it is predominantly the source of followed habit that finds these object preceded by "the".

While I have been quick to adopt the capitalization standard, I am still uncomfortable in most references that present these objects without "the" preceding them. Therefore, although from a technical standpoint I know I am incorrect, they are presented throughout my writings preceded by "the" as a result of followed habit.

Resources

Image Sources

SOHO Instrument Consortium
http://sohowww.nascom.nasa.gov
SOHO is a project of international cooperation between ESA and NASA

National Renewable Energy Lab
www.nrel.gov

Smithsonian Institution
The Dibner Collection

Scientific Data

Basic Nature by Andrew Scott
ISBN 0-631-17362-5

Physics by Hans C. Ohanian
ISBN 0-393-95401-3

Mapping the Sky by Leïla Haddad and Alain Cirou
ISBN 2-02-059692-x

The world wide web, various sites.

Keyword searches: physics, particle physics, nuclear radiation, thermodynamics, sun, passive solar, solar thermal, solar energy, solar system, etc.

Glossary

Absorption
The act or process of absorbing or the condition of being absorbed; the process in which incident radiated energy is retained without reflection or transmission on passing through a medium.

Alternative Energy
Energy derived from sources that do not use up natural resources or harm the environment.

Angle of Incidence
The angle that a ray of light falling on a surface or interface makes with the normal drawn at the point of incidence.

Antarctica
A continent largely within the Antarctic Circle (60 degrees South Latitude). Average temperatures across the entire land mass are below freezing, and the area is predominantly covered in ice. There are no permanent, human inhabitants and political sovereignty is suspended, in accord with the Antarctic Treaty.

Argon
A colorless, inert gaseous element constituting approximately one percent of Earth's atmosphere.Atomic number 18; atomic weight 39.948; melting point -372.7°F (189.3°C); boiling point -302.6°F (-185.9°C).

Aphelion
The point on the orbit of a celestial body that is farthest from the Sun.

Archimedes Screw
An ancient apparatus for raising water, consisting of either a spiral tube around an inclined axis or an inclined tube containing a tight-fitting, broad-threaded screw.

Atom
A unit of matter, the smallest unit of an element, having all the characteristics of that element and consisting of a dense, central, positively charged nucleus surrounded by a system of electrons.

Batt
Fiber wadded into rolls or sheets.

Btu
The quantity of heat required to raise the temperature of one pound of water from 60°F to 61°F at a constant pressure of one atmosphere.

Blackbody
A hypothetical object capable of absorbing all the electromagnetic radiation falling on it.

Glossary

Black Hole
A region of space resulting from the collapse of a star with a gravitational field so intense that its escape velocity is equal to or exceeds the speed of light.

Celsius
The Celsius thermometer or scale, so called from Anders Celsius, a Swedish astronomer who invented it, that registers the freezing point of water as 0° and the boiling point as 100° under normal atmospheric pressure.

Chromosphere
An incandescent, transparent layer of gas, primarily hydrogen, several thousand miles in depth, lying above and surrounding the photosphere of a star.

Compass
A device used to determine geographic direction, usually consisting of a magnetic needle or needles horizontally mounted or suspended and free to pivot until aligned with the earth's magnetic field.

Concrete
A synthetic, rock-like substance made by mixing sand and crushed stone with cement and water.

Conduction
A transfer of heat through matter by transfer of kinetic energy, from particle to particle.

Convection
The transfer of heat energy from place to place by the circulatory movement of a fluid mass. Usually restricted to liquids and gases.

Corona
The outermost region of the Sun's atmosphere; visible as a white halo during a solar eclipse A peculiar luminous appearance, or aureola, which surrounds the Sun, and which is seen only when the Sun is totally eclipsed by the moon.

Dark Ages
The period of history between classical antiquity and the Italian Renaissance.

Declination
The horizontal angle made by the magnetic needle of a compass in relation to the true north-and-south line of a given location as a result of the influence from Earth's magnetic field.

Density
The mass per unit volume of a substance at a specified pressure and temperature.

Desiccant
A substance, such as calcium oxide or silica gel, that has a high affinity for water and is used as a drying agent.

Direct Gain

A term used in passive solar design relating to solar heat collecting within the same space as where the collected heat will be utilized or stored.

Double Glazed

A term referring to the application of two panes of glass within a single window frame for the purpose of additional insulative value.

Drywall

A prefabricated material, such as gypsum based plasterboard used in building construction to line a wall or ceiling space.

Earthship

Dwellings whose primary building materials are used automobile tires, rammed full of earth, and waste aluminium cans. Earthships are typically built in an earth-bermed fashion or partially submerged.

Eave

The part of a roof that projects past the vertical plane of the walls upon which the roof rests. The primary purpose of an eave is to shed rainwater that collects on the roof beyond the plane of the walls, creating a weatherproofing mechanism for the structure.

Electromagnetic

Pertaining to or exhibiting magnetism produced by electric charge in motion.

Electron

A fundamental particle of matter. A constituent of the atom with a negative electrical charge.

Electroplate

To coat or cover with a thin layer of metal by electrode position.

Ellipse

An oval or oblong figure, bounded by a regular curve, which corresponds to an oblique projection of a circle, or an oblique section of a cone through its opposite sides. The greatest diameter of the ellipse is the major axis, and the least diameter is the minor axis.

Emission

The act of sending or throwing out; the act of sending forth or putting into circulation; issue; as, the emission of light from the Sun; the emission of heat from a fire.

Equator

The imaginary great circle around the earth's surface, equidistant from the poles and perpendicular to the earth's axis of rotation. It divides the earth into the Northern Hemisphere and the Southern Hemisphere.

Equilibrium

A condition in which all acting influences are cancelled by others, resulting in a stable, balanced, or unchanging system.

Glossary

Equinox

Either of the two times during a year when the Sun crosses the celestial equator and when the length of day and night are approximately equal; the vernal equinox or the autumnal equinox.

Fahrenheit

Of or relating to Fahrenheit's thermometric scale that registers the freezing point of water as 32° and the boiling point as 212° at one atmosphere of pressure.

Flat Plate Collector

A collector of solar heat designed to be flat.

Floorplan

A design layout showing the extents of the rooms and placement of walls and other significant elements within a house or building, as viewed from above.

Floor Slab

Generally referring to a floor space in a building created of concrete poured within a framework.

Flywheel

A regulator consisting of a heavy wheel that stores kinetic energy and smoothes the operation of a reciprocating engine.

Fossil Fuel

A hydrocarbon deposit, such as petroleum, coal, or natural gas, derived from living matter of a previous geologic time and used for fuel.

Foundation

A solid base, usually underground, constructed of either reinforced concrete or natural stone, upon which the walls of a building are raised.

Fundamental Particle

An particle that is less complex than an atom; regarded as constituents of all matter any of the particles of which matter and energy are composed or which mediate the fundamental forces of nature.

Fusion

The union of light atomic nuclei to form heavier ones under the conditions of extreme heat and pressure that results in a release of excess of energy. Fusion takes place in the core of stars, like the Sun, and has been replicated in Hydrogen bombs.

Gain

To acquire or obtain.

Glass

Materials with highly variable mechanical and optical properties that solidify from the molten state without crystallization; typically made by silicates fusing with boric oxide, aluminum oxide, or phosphorus pentoxide and are generally hard, brittle, and transparent or translucent; glass is considered to be a supercooled liquid rather than true solid.

Glazing

Glass set or made to be set in frames.

Gravity
The natural force of attraction between any two massive bodies, which is directly proportional to the product of their masses and inversely proportional to the square of the distance between them.

Gnomon
The style or pin, which by its shadow, shows the hour of the day. It is usually set parallel to the earth's axis. indicator provided by the stationary arm whose shadow indicates the time on the sundial.

Greyscale
A range of accurately known shades of grey.

Greenhouse Effect
A reference to the effect of increased heat in an enclosed environment due to the free introduction of electromagnetic radiation, which when subsequently transformed to heat cannot escape as readily as it enters.

Hydroelectric
Generating electricity by conversion of the energy of running water.

Imperial Measurement
A non-relational system of measurement. The units of mass are the pound and ounce, the units of measure are the inch, foot and yard; the units of volume are the ounce, quart and gallon.

Indirect Gain
A term used in passive solar design relating to solar heat collecting devices that collect solar heat in a space adjacent to where the heat will be utilized or stored.

Interference Pattern
A complex waveform produced when two or more waves interfere with one another.

Insulation
A material with properties that prevent the free passage of heat or electricity.

Isolated Gain
A term used in passive solar design relating to solar heat collecting devices with a remote position from where the collected heat will be utilized or stored.

Kelvin
The basic unit of thermodynamic temperature adopted under the International Units of Measure. A temperature scale in which zero occurs at absolute zero and each degree equals one kelvin. Water freezes at 273.15 K and boils at 373.15 K.

Kinetic Energy
The energy possessed by a body because of its motion, equal to one half the mass of the body times the square of its speed.

Latitude
The angular distance north or south of the Earth's equator, measured in degrees along a parallel, as on a map or globe.

Glossary

Lead
A soft, dense and malleable metal. Some of the chemical compounds found in lead are used in pigments such as printer's ink, pottery glazes and paint.

Lepton
Any of a family of elementary particles that participate in the weak interaction, including the electron and its associated neutrino.

Longitude
Angular distance on the earth's surface, measured east or west from the prime meridian at Greenwich, England, to the meridian passing through a position, expressed in degrees (or hours), minutes, and seconds.

Mass
A property of matter equal to the measure of an object's resistance to changes in either the speed or direction of its motion.

Mass Body
A term used to describe any volume of a material specifically designed to have a high degree of density in relation to its internal area.

Mass Wall
A term used to describe any volume of a material formed into a structural wall and specifically designed to have a high degree of density in relation to its internal area.

Meridian
An imaginary great circle on the earth's surface passing through the north and south geographic poles.

Metric System
A decimal-based system of measurement. The unit of mass is the kilogram, the unit of length is the metre, and the unit of volume is the litre.

Molecule
The smallest particle of a substance that retains the chemical and physical properties of the substance and is composed of two or more atoms; a group of like or different atoms held together by chemical forces.

Mould
A hollow container into which fluid or plastic material are put and allowed to harden or cure so as to take on the shape of the container.

Muon
An elementary particle not of the lepton family with a negative charge and a half-life of 2 microsecond; decays to electron and neutrino and antineutrino.

Neutrino
Any of three electrically neutral subatomic particles in the lepton family.

Neutron
A fundamental particle of matter. A constituent of an atomic core with a neutral electrical charge.

New Zealand
A country located in the South Pacific and a member of the British Commonwealth. Land mass is predominantly rolling hills to alpine, inclusive of active geothermal areas, plains, marshlands, rainforests, fiords and glaciers. Average regional temperatures vary widely, from 10 to 35 degrees C, due to the country's extended longitudinal orientation and varying local altitudes. Primary natural resources are timber, coal and natural gas. Industries include timber and processed timber products, meat, wool, dairy, fruit and vegetables, electrical goods, textiles and information technology.

Normal
Perpendicular to the direction of a tangent line to a curve or a tangent plane to a surface.

Oxidize
To combine with oxygen enter into a combination with oxygen or become converted into an oxide.

Parabola
A plane curve formed by the intersection of a right circular cone and a plane parallel to an element of the cone or by the locus of points equidistant from a fixed line and a fixed point not on the line.

Parabolic Trough
A square or rectangular reflective surface curved to form a point of focus along one if its axis.

Passive Solar
A construction-trades terminology referring to the practice of siting a home and constructing it's various elements so as to take advantage of the ability sunlight to provide heat.

Perihelion
The point nearest the Sun in the orbit of a planet or other celestial body.

Perpendicular
A line or plane falling at right angles on another line or surface, or making equal angles with it on each side.

Photon
The quantum of electromagnetic energy, regarded as a discrete particle having zero mass, no electric charge, and an indefinitely long lifetime.

Photosphere
The visible, intensely luminous surface of a star.

Photovoltaic
Capable of producing a voltage when exposed to radiant energy, especially light.

Physics
The science of matter and energy and of interactions between the two, grouped in traditional fields such as acoustics, optics, mechanics, thermodynamics, and electromagnetism, as well as in modern extensions including atomic and nuclear physics, cryogenics, solid-state physics, particle physics, and plasma physics.

Glossary

Pi

A transcendental number, approximately 3.14159, represented by the symbol π which expresses the ratio of the circumference to the diameter of a circle and appears as a constant in many mathematical expressions.

Plasma

An electrically neutral, highly ionized gas composed of ions, electrons, and neutral particles. It is a phase of matter distinct from solids, liquids, and normal gases.

Plywood

A structural material made of layers of wood glued together, usually with the grains of adjoining layers at right angles to each other.

Polymer

Any of numerous natural and synthetic compounds of usually high molecular weight consisting of up to millions of repeated linked units, each a relatively light and simple molecule.

Polystyrene

A rigid thermoplastic used for the construction of containers and other moulded products requiring insulative properties.

Polythene

A lightweight thermoplastic; used especially in packaging and insulation.

Potential Energy

The energy that exists in a body as a result of its position or condition rather than of its motion.

Power Tower

A system for generating power from sunlight that uses a collection of mirrors that reflect sunlight to a central light-receiving station.

Proton

A fundamental particle of matter. A constituent of an atom with a positive electrical charge.

Quark

Any of a group of six elementary particles having electric charges of a magnitude one-third or two-thirds that of the electron.

R-value

A measure of the capacity of a material, such as insulation, to impede heat flow, with increasing values indicating a greater capacity measured in British Thermal Units (Btu).

Radiation

Energy released by the emission of waves or particles.

Reflector

A polished surface for reflecting light or other radiation.

Reichstag

The Diet, or House of Representatives, of pre-1945 Germany, which was composed of members elected for a term of three years by the direct vote of the people.

RSI
A measure of the capacity of a material, such as insulation, to impede heat flow, with increasing values indicating a greater capacity measured in the International Standard Units of Measure (metric).

Selective Surface
Surface coating materials that are known to produce a specific heat collecting/emitting characteristic.

Slipform
A large type of mould commonly used in rammed earth and concrete construction. The slipform has been adapted for use in stone masonry construction.

Smithsonian
Of or pertaining to the Englishman J. L. M. Smithson, or to the national institution of learning, which he endowed at Washington, D. C., as the Smithsonian Institution.

Soffit
The area underneath the over-hanging portion of a roof.

Solar
Of, relating to, or proceeding from the Sun

Solar Chimney
The arrangement of a chimney structure in relation to a building so that the chimney may be used to draw air from within the building as a result of natural convective processes.

Solar Flare
A sudden eruption of hydrogen gas on the surface of the Sun.

Solar Roof
A design feature of a building that uses mass materials in or on its roof to collect solar heat.

Solar System
The Sun together with the nine planets and all other celestial bodies that orbit the Sun.

Solar Thermal
Heat obtained from the Sun.

Solar Wind
A stream of high-speed, ionized particles ejected primarily from the Sun's corona.

Solstice
Either of two times of the year when the Sun is at its greatest distance from the celestial equator. The summer solstice in the Northern Hemisphere occurs about June 21, when the Sun is in the zenith at the tropic of Cancer; the winter solstice occurs about December 21, when the Sun is over the tropic of Capricorn. The summer solstice is the longest day of the year and the winter solstice is the shortest.

Spacetime
The four-dimensional continuum of one temporal and three spatial coordinates in which any event or physical object is located.

Glossary

Strong Force

A terminology from the practitioners of particle physics, that describes the bond holding together protons and neutrons in an atomic nucleus.

Sun

A star that is the basis of the solar system and that sustains life on Earth, being the source of heat and light.

Sunlight

The light produced as a result of the nuclear fusion processes occurring within the Sun.

Sunspot

Any of the relatively cool dark spots appearing periodically in groups on the surface of the Sun that are associated with strong magnetic fields.

Supernova

A rare celestial phenomenon involving the explosion of most of the material in a star, resulting in an extremely bright, short-lived object that emits vast amounts of energy.

System

A group of interacting, interrelated, or interdependent elements forming a complex whole.

Temperature

A measure of the average kinetic energy of the particles in a sample of matter, expressed in terms of units or degrees designated on a standard scale.

Thermal Diffusivity

The speed of temperature change in a material when it is exposed to a temperature change in its environment; a measure of heat flowing freely through a material.

Thermodynamics

The branch of physics concerned with the conversion of different forms of energy.

Thermometer

An instrument for measuring temperature, especially one having a graduated glass tube with a bulb containing a liquid, typically mercury or colored alcohol, that expands and rises in the tube as the temperature increases.

Thermosiphon

A naturally occurring phenomenon whereby cold air or fluid is drawn into a space as a result of convective processes.

Trombe Wall

A passive solar heating device, invented by French scientist Felix Trobme, that uses massive walls, vents and glass covering to achieve it's heating effect.

U-value

A measure of energy passing through a volue of material in an hour, for every degree of temperature difference across the material.

Vacuum

A space absolutely devoid of matter.

Water

A clear, colorless, odorless, tasteless and very slightly compressible liquid oxide of hydrogen H_2O which appears bluish in thick layers; essential for most plant and animal life and the most widely used of all solvents. Freezing point 0°C (32°F); boiling point 100°C (212°F); specific gravity (@4°C) 1.0000; weight per gallon (@15°C) 8.338 pounds (3.782 KG). One of the most effective natural heat storage mediums.

Glossary

Index

A

Adams, William 52

Africa 56, 60, 61

Algeria 50, 174

alternative energy 59, 145, 185

aluminum 21, 102, 152, 188
 foil 21

ammonia 56, 175, 177

angle of incidence 24, 26, 27, 34, 35, 38, 150, 152, 185

Antarctica i, 35, 159, 185

aphelion 22, 185

Archduke Ferdinand 61

Archimedes 21, 41, 42, 63, 172, 185

argon 127, 185

astrophysicist 48, 68, 77

atmosphere 34, 38, 74, 82, 175, 185

atom 77, 80, 81, 82, 84, 85, 90, 92

atomic 70, 77

aurora
 australis 74
 borealis 74

B

blackbody radiation 90

black hole 70

boiler 4, 175, 176, 179

Btu 185, 192

C

Celsius 69, 179, 186

charge
 negative 81, 190
 positive 81
Chromosphere 186

compass 186

compound 79, 83, 175

concrete 186

conduction 94, 186

Conservation of Mass, Law of 46

Convection 94, 95, 186

corona 73, 74, 75, 193

D

Da Vinci, Leonardo 173

Density 186

De Saussure, Horace 174

differential rotation 72

E

Earth 4, 9, 67, 70, 74, 81, 82, 84, 86, 92, 183, 185, 186, 189, 194

Earthship 3, 4, 187

Egypt 61, 176

Einstein, Albert 67, 119

electromagnetic 82, 83, 84, 90, 91, 117, 118, 121, 185, 189, 191

electron 80, 81, 189, 190, 192

electroplate 187

element 79, 80, 82, 162, 185, 191

emission 76, 122, 123, 124, 127, 187, 192

emit 122, 123, 125, 127

Aubrey Eneas 175, 176

energy
  kinetic 86, 88, 90, 121, 186, 188, 194
  potential 192

epoxy 124

equator 36, 72, 187, 189, 193

equilibrium 187

equinox 30, 187

Ericsson, John 54

ether 60

Euclid 41

F

Fahrenheit 46, 69, 107, 112, 187

fiberglass 110

fixed collector 24, 28, 35, 38

flat plate collector 56, 60

floor slab 161, 162, 163, 164, 165

fluid 16, 18, 19, 20, 42, 56, 60, 94, 103, 104, 123, 178, 186, 190, 194

fossil fuel 188

France 50, 56, 57, 179, 182

freon 106

frequency 117, 118

fundamental particle 187, 190, 192

fusion 81, 82, 188

G

gain
  direct 14, 15, 186
  indirect 16, 189
  isolated 18, 189
gamma rays 84, 118

gas 19, 20, 56, 59, 72, 90, 94, 106, 127, 178, 185, 186, 188, 191, 192, 193

glass 16, 17, 18, 19, 20, 21, 44, 48, 50, 52, 56, 60, 93, 103, 109, 115, 127, 137, 145, 156, 157, 168, 173, 174, 175, 187, 188, 194

gnomon 28, 188

gold 123, 124

graphite 122, 123

grayscale 119

greenhouse 44, 48, 145

Greenwich, England 33, 190

H

heat 9, 10, 14, 16, 17, 18, 19, 20, 21, 22, 28, 34, 41, 44, 48, 50, 52, 54, 56, 59, 60, 62, 63, 67, 74, 77, 82, 83, 84, 85, 87, 90, 92, 93, 94, 95, 96, 99, 101, 102, 103, 104, 105, 106, 107, 108, 110, 112, 113, 114, 115, 117, 119, 121, 122, 123, 124, 125, 127, 128, 133, 137, 142, 153, 154, 156, 157, 158, 161, 163, 171, 173, 174, 179, 180, 181, 185, 186, 187, 188, 189, 191, 192, 193, 194, 195

Herschel, Sir John 48, 49, 174

Hertzsprung, Ejnar 68, 71

horizon 22, 24, 35

horizontal orientation 24

hydroelectric 1

I

India 52, 56, 175, 180, 182

infrared 118, 121, 122, 123, 127

insulation 4, 5, 18, 99, 107, 109, 110, 112, 114, 115, 145, 157, 161, 162, 164, 165, 166, 174, 192

interference pattern 118

internal energy 85, 87, 88

K

Kelvin 68, 69, 189

Keppler, Johannes 22

L

lacquer 124

Langley, Samuel Pierpoint 48, 49, 50, 175

La Voiser, Antoine 46, 47, 174

light 10, 21, 34, 41, 44, 50, 54, 63, 68, 70, 73, 74, 78, 84, 95, 108, 114, 117, 118, 119, 121, 122, 123, 125, 137, 168, 172, 174, 185, 186, 187, 188, 191, 192, 194

longitude 32, 34, 190

M

mass body 14, 19

mass wall 154, 156, 157

materials vi, 1, 2, 9, 10, 13, 14, 16, 18, 21, 46, 47, 62, 63, 86, 95, 96, 99, 101, 102, 103, 104, 105, 106, 107, 108, 109, 110, 111, 112, 114, 115, 117, 119, 122, 123, 124, 127, 128, 133, 138, 144, 153, 156, 165, 168, 187, 193

Maxwell, James Kirk 90

McMurdo 1, 2

MDF 145

microwave 81

midwinter 37, 160

Milky Way 68, 72, 172

mineral oil 105

mirror 173, 175

molecule 190

Mouchot, August Bernard 50, 51, 52, 54, 56, 58, 59, 124, 142, 174, 175

Mt. Whitney 48, 50

Muon 190

N

nanometer 118

Napoleon III 50

nature 18, 44, 58, 60, 82, 88, 90, 104, 117, 118, 119, 124, 146, 168, 188

neutrino 189, 190

neutron 190

Newton, Sir Isaac 90, 173

New Zealand i, vi, 7, 159, 168, 181, 182, 190

northern hemisphere 150, 160, 168, 187, 193

north pole 32

nuclear vii, 77, 79, 81, 82, 83

nucleus 77, 81, 82, 83, 84, 185, 193

O

orientation 24, 25, 32, 34, 35, 36, 37, 38, 152, 160, 190

oxidation 124, 125, 140

oxide 186, 188, 191, 195

oxidize 124

P

pacific ocean 159

parabolic trough 21, 54, 58, 60, 175

particle 67, 79, 81, 119, 184, 186, 187, 188, 190, 191, 192, 193

passive solar 6, 10, 27, 41, 67, 153, 154, 155, 158, 159, 160, 161, 168, 172, 177, 182, 184, 186, 189, 194

perihelion 191

perpendicular 26, 191

phase-change 106

photosphere 72, 73, 74, 186

photovoltaic 149, 191

physics 184, 191

Planck, Max 90, 91, 92, 119, 122

polar 28, 31, 34, 38, 74

polystyrene 4, 109, 111, 113, 143, 145, 161, 162, 163, 165, 166, 167
  expanded 110
  extruded 109, 162

powder coating 124

Power Tower 52, 53, 54, 175, 180, 181, 192

prime meridian 32, 33, 190

proton 192

Q

quantum mechanics 119

quark 192

R

R-value 107, 108, 109, 110, 111, 112, 113, 192

radiate 103

radiating 16, 76, 90, 157

radiation 28, 35, 44, 54, 67, 76, 77, 80, 82, 83, 84, 85, 90, 91, 92, 93, 94, 114, 117, 119, 121, 122, 123, 124, 125, 127, 153, 156, 162, 178, 184, 185, 189, 192

ray 26, 73, 185

reflector 138, 192

Reichstag 61, 192

relativity 67

right angle 26

RSI 107, 108, 109, 110, 192

Russell, Henry Norris 68, 71

S

selective surface 122, 123, 124

Shuman, Frank 54, 60, 61, 62, 176

silicon 123, 124

slipform 158, 193

Socrates 41, 172

soffit 193

Sol 68

Solar
  chimney 193
  designing 9, 10, 26, 77, 90, 128, 133, 171
  flare 74
  furnace 46, 47, 173, 174, 178
  heat 10, 14, 21, 22, 34, 52, 56, 59, 63, 67, 84, 99, 101, 103, 104, 105, 110, 119, 122, 128, 133, 154, 173, 186, 188, 189, 193
  Motor 50
  oven 4, 5, 144, 145, 176, 177, 178
  prominence 75
  radiation 28, 35, 54, 67, 76, 77, 80, 84, 85, 90, 93, 94, 119, 124, 125, 127, 153, 156, 162
  roof 193
  thermal vii, 13, 15, 17, 21, 23, 25, 27, 29, 31, 33, 35, 37, 39, 142, 193
  tracking viii, 149
  wind 74
  window 6, 7

Solstice 31, 193

Southern Hemisphere 168, 187

South Pole 1, 2, 4, 32, 36

spacetime 22, 32

spectrum 117, 118, 121

stainless steel 21, 105, 123, 140

steam 44, 50, 51, 52, 53, 54, 58, 60, 123, 174, 175, 176, 177, 179

steam engine 50, 52, 54, 60, 174, 175, 177

Stone House iv, 158

strong force 77, 82, 84

Styrofoam 109

sulphur dioxide 59

summer 3, 22, 23, 30, 35, 38, 135, 150, 152, 154, 158, 162, 164, 168, 172, 193

Sun vii, viii, 4, 5, 9, 10, 13, 14, 16, 22, 23, 24, 25, 26, 27, 28, 29, 30, 32, 34, 35, 37, 38, 39, 41, 44, 46, 48, 50, 52, 54, 56, 60, 61, 63, 67, 68, 69, 70, 71, 72, 73, 74, 75, 76, 77, 78, 81, 82, 83, 84, 90, 92, 101, 103, 104, 105, 117, 122, 123, 133, 135, 137, 140, 143, 144, 149, 150, 151, 152, 154, 155, 158, 159, 160, 162, 168, 171, 172, 173, 174, 179, 182, 183, 185, 186,

187, 188, 191, 193, 194

sunlight 4, 9, 10, 14, 16, 18, 19, 20, 21, 26, 27, 28, 34, 38, 41, 43, 44, 45, 46, 48, 52, 53, 58, 59, 60, 62, 63, 67, 68, 78, 84, 90, 92, 96, 103, 104, 109, 117, 121, 123, 127, 137, 142, 144, 150, 154, 156, 160, 170, 173, 174, 179, 180, 191, 192

Sunrise 159

sunset 159, 160

sunspot 194

supernova 194

sustainable 9, 81, 181

T

tea viii, 135

Tellier, Charles 56, 57, 59, 60, 175

thermal conductivity 101, 102

thermal diffusivity 102, 194

thermal flywheel 156, 162

thermal storage 14, 156

thermodynamics
  first law of 87
  second law of 92
  zeroth law of 93

thermodynamic system 86

thermometer 4, 6, 108, 186

thermosiphon 16, 17, 95, 156

Trombe, Felix 6, 16, 156, 157, 175, 177, 178, 194

truncated cone 50, 51

turbine 53, 178

U

U-value 107, 112, 113, 127, 171, 194

ultraviolet 118, 168

V

vertical orientation 24, 25, 35, 37, 38, 152

Volumetric Heat Capacity 102, 194

W

warmth 16, 35, 41, 45, 63, 84, 103, 158, 182

water 14, 15, 16, 18, 21, 34, 44, 48, 50, 53, 56, 58, 59, 60, 61, 62, 63, 69, 85, 95, 99, 103, 104, 105, 106, 114, 122, 123, 124, 138, 140, 154, 156, 162, 165, 167, 168, 172, 173, 174, 175, 176, 177, 178, 179, 180, 181, 182, 185, 186, 187, 189

wave 67, 117, 118, 119

wave-particle duality 67, 119

wavelength 117, 121, 122

Wave Theory of Light 119

Western Hemisphere 33

white dwarf 70

Willsie, Henry E. 59, 60, 176

winter 3, 22, 23, 24, 30, 34, 35, 37, 38, 135, 150, 152, 153, 156, 158, 164, 168, 170, 193